PRAISE FOR *Restoring Heritage Grains*

"Eli Rogosa has delivered to us, her many fans, the long-awaited book, *Restoring Heritage Grains*, in which she totally blows the lid off of this historic moment in the world of bread. She not only artfully guides us through thousands of years of the history and botanical evolution of wheat but also, prophetically, shows us its very future. And now we all have access to Eli's inner world, to the passion that has been fermenting within her for many years and now exists forever through her brilliant words."

— Peter Reinhart, educator; author of *Bread Revolution*

"*Restoring Heritage Grains* offers a veritable treasure trove from the past, yet one that is very relevant for today! The book introduces truly healthier, more nutritious, beautiful, and exciting grains to cultivate in your garden and farm and to enhance your palate. Read, grow, preserve, eat, and enjoy ancient grains for a biodiversity of taste and nourishment!"

— John Jeavons, author of *How to Grow More Vegetables*; executive director of Ecology Action

"Eli Rogosa has lived among the world's few remaining peasant farmers who continue to cultivate landrace wheat seeds and traditions. She has collected and faithfully tended and multiplied their unique local varieties, learned their traditional production techniques, and recorded their special recipes. She brought them to her home in New England and crossed them to combine their qualities and adapt them to the very different climate of their new home. Now, in *Restoring Heritage Grains*, she shares the wealth of information that she has preserved and the flavor of the seeds that she has saved, with people in this country and around the world."

— Klaas Martens, farmer, Lakeview Organic Grain, Penn Yan, NY

"This beautiful book is unlike any other publication on wheat or grains that I have ever read. Written poetically, it is a rare mix of science, history, and culture; therefore, the book will be equally inspiring for scientists, students, farmers, seed savers, culinary experts, or just any person looking for interesting reading. With this book, Eli gives us a key to restoring our bread of life."

— Mariam Jorjadze, director, Biological Farming Association Elkana (Georgia)

"Let yourself be inspired by the inflammable enthusiasm of Eli Rogosa about the diversity of ancient wheats, their historical backgrounds, and notes from her many encounters in different countries. The author brings these wheats not only into your stomach with lots of recipes, but also into your heart, which is the most important step on their way into the fields, where they can develop in our modern times into what wheat should be for humans: a well-balanced partner that can help us to cultivate our minds, our bodies, and our sentiments."

— Dr. Karl-Josef Mueller, biodynamic cereal breeder
at Cereal Breeding Research, Neu Darchau, Germany

"Eli Rogosa deserves credit for pioneering the current return of interest in heritage grains. In a compelling and inspiring book, she retraces her own voyage of discovery into the beauty and importance of endangered grain varieties, the tragic loss of their presence in our fields and diets, and how we can participate in returning this most ancient of foods to our tables. Her wide-ranging work is a powerful reminder of the depth of our connection to the first crops cultivated by humans."

— Sylvia Davatz, Solstice Seeds

"In this book, agro-anthropologist, farmer, and baker Eli Rogosa helps us rediscover ancient landrace and traditional pre-Green Revolution wheats — varieties that are more delicious, nutritious, drought-resistant, and resilient than modern wheats, and that are already organic-adapted. The author covers everything from the romantic to the practical: personal stories about finding individual plants of rare wheats in Israel; historical and anthropological information; methods for growing, harvesting, and threshing; as well as many detailed recipes. A must read for anyone who has a garden or farm and who likes good bread."

— Carol Deppe, author of *The Tao of Vegetable Gardening*

"Our common cultural history goes all the way back to the very roots of civilization: the domestication of the cereals 12,000 years ago. In page after page of this book, Eli Rogosa's profound knowledge, love, and passion for our common culinary and genetic heritage links our history with our daily bread, and fills the reader with enthusiasm to go into the field, and into the kitchen, to follow her example: Grow it, bake it, and eat it! Eli Rogosa's quest for restoring quality bread from heritage grains is not only for the sake of your own health but to restore what unites us all, and thereby a mission of peace."

— Dr. Anders Borgen, organic wheat breeder, Denmark

"Most wheat grown worldwide today can be described as an in-bred, dwarfed, distant cousin of the genetically diverse, farmers' landrace cereal crops of the past. Eli Rogosa argues passionately and convincingly in her book that from many perspectives, including food security and nutritional value, our landrace cereals need to be brought back from the brink of extinction. Eli illustrates the central role of cereals in human civilization as we know it, including in myth and religion and how this role has been traduced by agribusiness interests. Eli adds valuable advice and knowledge for the grower and the cook on preservation and use of our cereal crop inheritance."

— Andy Forbes, secretary, Brockwell Bake Association, London, UK

"*Restoring Heritage Grains* is both poetic and practical. Eli Rogosa first tells the sad story of how the Green Revolution transformed the staff of life into a toxic-drenched monocrop. Then she shares the joyful story of her life's work discovering, growing, distributing the seed and spreading the word about heritage grains. She makes a compelling case for heirloom landraces, the deep-rooted, diverse gene pools that coevolve with changing conditions, "people and seeds" finding ways to survive through climate challenges. . . . This is a book to cherish."

— Elizabeth Henderson, author of *Sharing the Harvest*

"This is a marvelous book, which I will read again and again over the years. Eli has woven a tapestry of fact and flavour, drawing on botanical, agricultural, nutritional, and folk information never before assembled under one cover. And she has included practical information on how to make delicious bread and beer. She has described how the first farmers were 'evolutionary plant breeders' and worked with nature to create the biodiverse crops we now call 'heritage' grains. . . . This book is a critique of industrial agriculture, but it is also a practical manual for how to reintroduce diversity into our farming systems by growing heritage grains, and how we can help repair our spiritual relationship with the earth."

— John Letts, archaeo-botanist and farmer, Heritage Harvest Ltd., Oxford, UK

Restoring Heritage Grains

The Culture, Biodiversity, Resilience, *and* Cuisine *of* Ancient Wheats

ELI ROGOSA

Chelsea Green Publishing
White River Junction, Vermont

Project Editor: Benjamin Watson
Copy Editor: Ben Gleason
Proofreader: Eileen M. Clawson
Indexer: Shana Milkie
Designer: Melissa Jacobson

The top front cover image shows, *left to right*, black winter emmer, Banatka, North African black beard durum, Rouge de Bordeaux, and einkorn. Photograph by Amy Toensing. Please see page 7 of the color insert for a full description of the bottom front cover image.

Printed in the United States of America.
First printing June, 2016.
10 9 8 7 6 5 4 3 2 1   16 17 18 19 20

Our Commitment to Green Publishing
Chelsea Green sees publishing as a tool for cultural change and ecological stewardship. We strive to align our book manufacturing practices with our editorial mission and to reduce the impact of our business enterprise in the environment. We print our books and catalogs on chlorine-free recycled paper, using vegetable-based inks whenever possible. This book may cost slightly more because it was printed on paper that contains recycled fiber, and we hope you'll agree that it's worth it. Chelsea Green is a member of the Green Press Initiative (www.greenpressinitiative.org), a nonprofit coalition of publishers, manufacturers, and authors working to protect the world's endangered forests and conserve natural resources. *Restoring Heritage Grains* was printed on paper supplied by Thomson-Shore that contains 100% postconsumer recycled fiber.

Library of Congress Cataloging-in-Publication Data
Names: Rogosa, Eli, 1952– author.
Title: Restoring heritage grains : the culture, diversity, resilience, and
   cuisine of ancient wheats / Eli Rogosa.
Description: White River Junction, Vermont : Chelsea Green Publishing, [2016]
   | Includes bibliographical references and index.
Identifiers: LCCN 2016010222 | ISBN 9781603586702 (pbk.) | ISBN 9781603586719 (ebook)
Subjects: LCSH: Wheat — Heirloom varieties.
Classification: LCC SB191.W5 R64 2016 | DDC 633.1/1 – dc23
LC record available at http://lccn.loc.gov/2016010222

Chelsea Green Publishing
85 North Main Street, Suite 120
White River Junction, VT 05001
(802) 295-6300
www.chelseagreen.com

MIX
Paper from
responsible sources
FSC
www.fsc.org   FSC® C013483

Contents

CHAPTER 1

On the Verge of Extinction

Wheat whispers the journeys of the people who planted them: the village traditions, the trading and migrations that are kneaded into our breads. The heritage wheat of North America originated in the majestic landraces of the Fertile Crescent and Old Europe. When people immigrated to the New World, they brought cherished landrace seeds from their homeland. These are the wheats that nourished earlier peoples, but today they are almost lost, replaced by modern "Green Revolution" wheat dependent on agrochemicals to survive.

For the past 12,000 years farmers have selected seed, generation by generation, to develop the landrace[1] wheats that nourished civilizations. Landrace wheats have robust root systems that reach out to absorb organic nutrients, height that shades out encroaching weeds, root exudates that suppress the weeds (no herbicides needed), complex resistances to local diseases, and incomparable flavor. Yet who today has heard of a landrace?

Today genetic management of wheat has shifted to the hands of industrial breeders who have replaced landraces worldwide with patented commercial varieties that are bred for uniformity in agrochemical-soaked fields. Modern wheats' narrow genetic base leaves them vulnerable to disease and lacking adaptability to the weather extremes of drought and heavy rain. Global warming looms menacingly. Industrial wheat, the most widely grown crop on the planet, is a teetering monocrop that has been fine-tuned for predictable weather in computer-controlled mega-farms; dwarfed for efficient harvest; and dependent on irrigation, pesticides, and synthetic fertilizers to survive. It is bred like a pedigree racehorse for high performance in optimal conditions, but it is weak under adversity. Just as a monocropped variety of potato was wiped out in Ireland by a strain of blight in 1845, the uniformity of modern wheat is a disaster waiting to happen.

Economies of scale enable industrially bred wheat to be produced cheaply and yield well in favorable conditions, but with hidden costs. The industrial food system leaves 1.3 billion people hungry worldwide, yet it uses more fertilizer, more pesticides, and more energy than low-input ecological agriculture. Modern wheat has been bred to be addicted to a fix of high-nitrogen fertilizers. But what did the peasants of yore grow? Where are the almost-forgotten traditional varieties?

This book tells the story of my journey to explore the history and heritage of wheat. We have been scammed by multinational corporations to the extent that few US farmers today even know what landrace wheat is. After years of research I want to shout the truth to the world: that the almost-forgotten heritage wheats have higher yields than modern wheat in organic soils. That they are safer for people with gluten allergies. That we have been robbed of our inheritance of seed saving. Wheat is easy to grow. Small-scale farmers and gardeners can be self-sufficient in growing our own grains.

Modern wheat is no longer nourishing us. It is an empty harvest bloated with nitrate chemicals, causing an epidemic of obesity and ill health. Modern breeding has been so successful in achieving its goal of covering the world with uniform Green Revolution varieties that the heritage wheats that were once grown by traditional farmers worldwide are disappearing. Few heritage wheats are grown in the United States, and they are minimally available worldwide. The unprecedented erosion of wheat biodiversity threatens not only food security, but also nutrition and culinary art.

Even the Fertile Crescent countries of Israel, Palestine, and Jordan, the ancestral home of wild wheat, today buy over 95 percent of their wheat from US mega-farms. The most delicious wheats — the vast biodiversity of wheat species that are best adapted to organic fields, the wheats that nourished civilizations — are almost lost to the world.

The story of how humankind's staple food has become a toxic substance reveals wheat as an indicator organism, like a canary in a coal mine. Indeed, it is because of the central, sacred role of the ancient "staff of life" that the inner realities of modern agriculture, world trade, and industrial food are dramatically exposed. Can we reach into the heart of this beast and restore wheat into the majestic nourishing being it once was? The magnificent health and rich diversity of landrace wheat is transformative, imparting hope to each of us. Like Gandhi's Salt March, when he led the people to collect free life-giving salt, each of us has the potential to feed ourselves and our communities using the restoration of humankind's age-old, staple food crop as an inspiration.

The Roots of Modern Wheat

The scientific revolution of the nineteenth century heralded the rise of reason. Charles Darwin formulated the theory of evolution. Gregor Mendel developed the theory of inheritance, making possible a new approach to breed single plants with specific traits. This method, known as pedigree breeding, creates uniform pure lines with little capacity for adaptation.

In 1868 a Scottish breeder found a unique plant in his field that was shorter and stockier than other wheats. The stalk was exceptionally sturdy. He selected this one plant and multiplied it. Its stocky seedhead suggested the name of "Squarehead." It became popular as the increased use of synthetic nitrogen fertilizer spread. Soon, much of the wheat planted in Europe became crossed with Squarehead. Landraces disappeared throughout Europe.

Norman Borlaug, father of the Green Revolution, was a hard-working Midwestern farm boy who came of age in the grueling years of the Depression. Knowing well the pangs of hunger, he devoted his life to alleviating starvation by breeding high-yielding wheats that were dwarfed to not "lodge" (fall over) in soils fed by synthetic nitrogen fertilizer. Borlaug combined diverse genes for complex disease resistance, while streamlining the wheats for similar heights, flowering times, and maturity dates to advance conventional large-scale farming systems the world over. Borlaug's brilliance developed under the specter of "progress through chemicals." Chemical and agribusiness industries profited from Borlaug's promotion of their products of patented seeds and herbicides.

To overcome the problem that dwarfed wheats are towered over by normal-height weeds, Borlaug's "miracle wheat" relied on the heavy use of agrochemicals. Borlaug believed that "synthetic fertilizer only replaces substances naturally present in the soils anyway."[2] He did not notice that the loss of the beneficial disease-controlling bacteria, mycorrhizal fungi, and earthworms invited susceptibility to attack from insects and pathogens. Although Borlaug's dwarfed wheats do yield higher than traditional varieties when given intensive irrigation, fertilizers, and pesticides, the traditional varieties outperform the modern varieties in organic farming systems. The dwarfed wheats fail under the stresses of weather extremes, whereas einkorn, emmer, and many other landrace wheats thrive. I documented this in four years of scientific trials funded by the USDA Sustainable Agriculture Research and Education Program.

THE LOSS OF BIODIVERSITY AND INDIGENOUS KNOWLEDGE

We are living in a period of unprecedented extinctions. The UN Food and Agriculture Organization (FAO) estimates that 90 percent of the food crops grown 100 years ago have disappeared from farmers' fields and are functionally extinct.[3] Of the approximately 250,000 plant species on our planet, about 50,000 are edible, yet only 15 crops provide 90 percent of our calories. Wheat, corn, and rice provide 60 percent, yet only a tiny fraction of their vast biodiversity is grown in the modern field. Since wheat is the most widely grown food crop, modern plant breeders have focused on it more than any other crop. Industry's obsession with genetic uniformity, high yield, and profit has taken the work of Borlaug to a dangerous extreme by replacing the world's wealth of biodiversity with monocultures inconceivable a generation ago.

Modern wheat systems not only deplete the self-regenerating cycles of living soil but have destroyed the indigenous farming knowledge of generations of peasant farmers. The primary goal of industrial breeders is to increase yield, broadening adaptability and productivity over as wide an area as possible. The uniform standardization of varieties legally protects "plant breeder's rights" so breeders can claim royalties. These imperatives promote genetic homogeneity, reduce variables in the growing environment, and eliminate the traditional seed mixtures that foster greater adaptability to local environments through population diversity.

When biodiversity is lost, the capacity of a crop to evolve and adapt is destroyed. We do not have to wait until the last landrace wheat dies before wheat is on the verge of extinction. It becomes threatened when it loses its ability to evolve and adapt to new conditions, when neither its bottlenecked genetics nor its load of chemicals can protect it. That day has come quietly as modern wheat blankets the earth.

Before the introduction of Green Revolution monocultures, wheat was intercropped and rotated with legumes and other diverse soil-building crops. Heavy infusions of chemical fertilizers destroy the vital soil organisms that are nature's recycling system for nutrients and the soil's natural defense against disease and pathogens. The global wheat system's agrochemicals have sterilized millions of acres of formerly fertile topsoil. Ironically, Borlaug's wheats, created to feed the world, have set the stage for the depletion of the world's agricultural soils.

FOLLOW THE MONEY

With the multiple, problematic side effects of chemical-dependent modern wheat, why then would intelligent scientists introduce it? Let us follow the chain of actors to discover who really benefits from the Green Revolution. In low-input organic wheat fields the world over, landrace wheats outperform modern wheat, so why would anyone want to phase them out? Who benefits from the petroleum-based wheat system? You guessed it. The energy and oil companies are the ones who reap vast profits. Who is behind these huge energy corporations?

The Rockefellers' oil empire began in 1870 when John D. Rockefeller founded Standard Oil. He became America's first billionaire, and later the richest man in the world. After the US Supreme Court created antitrust laws in 1910, ruling that monopolies were illegal, Standard morphed into smaller interlinked companies. The successor companies from Standard Oil's breakup form the core of today's US oil industry that includes ExxonMobil, British Petroleum (BP), Chevron, and more. The Green Revolution was the brainchild of the Rockefeller Foundation in partnership with Monsanto, Cargill, and DuPont. Green Revolution wheat increases wheat productivity with petrochemicals at a huge environmental and health cost, while consolidating economic profit into the hands of an elite few.

The Green Revolution's dramatic increase in wheat yield created a pressing need to exploit new world markets. Tragically, the subsidized wheat sold to developing countries not only replaced landrace wheats but put Third World farmers out of business.

Since the 1960s, the Jordanian government has bought subsidized US wheat that is sold at below-market prices throughout the country. Who benefits? The Jordanian national seed bank director explained the full picture to me. He was livid and felt powerless. While US-based multinational wheat companies make huge profits, the small-scale Jordanian farmers cannot compete with subsidized US wheat. Green Revolution seeds did not prevent a famine in Jordan. They caused it. Precious Jordanian landrace wheats are on the verge of extinction as a result.

Vandana Shiva reports how the Indian government subsidized Indian wheat for the global market, only to hike prices for their own people, causing widespread hunger at home.

Iraqi farmers have been saving wheat and barley seed since the dawn of agriculture. However, as part of economic restructuring by the Bush administration, Iraqi farmers are now forbidden to save their own seeds. The George W. Bush administration updated Iraq's intellectual property law to "meet current international standards of plant protection," making it illegal to save the irreplaceable drought-hardy landrace seeds that evolved over millennia in the hands of traditional peasants.[4] Farmers are forced to buy patented modern seeds from Monsanto, Cargill, and the World Wide Wheat Company.

EATING OIL: THE GREEN REVOLUTION FEEDS GLOBAL WARMING

Modern wheat is dependent on petroleum. Green Revolution foods are bred with an umbilical cord to fossil fuel–based chemical fertilizers, pesticides, and herbicides, and to an oil-based supply chain. A decline in oil and gas production would cause catastrophic famines. At a time when we should be decreasing greenhouse emissions, the food system is increasing its carbon footprint to the point where it has become a significant contributor to global warming. While Green Revolution cereal production has more than doubled in developing regions such as India, Africa, and Asia, the ratio of crops produced to energy input has decreased. Vast amounts of oil and gas are used as raw materials and energy in the manufacture of fertilizers and pesticides for all aspects of production, from planting, irrigation, feeding, and harvesting to processing, distribution, and packaging. Fossil fuels are essential in the construction and the repair of farm machinery, processing facilities, storage, ships, trucks, and roads. The global wheat supply system is one of the biggest consumers of fossil fuels and one of the greatest producers of greenhouse gases.

Chemical fertilizers account for at least 38 percent of the greenhouse gas emissions from agriculture. They emit nitrogen oxide, which is 300 times more potent than carbon dioxide as a greenhouse gas.[5] Removing chemical fertilizers is a critical strategy that can reduce the emissions that cause climate change. Like the addiction to oil, the addiction to chemical fertilizers ultimately benefits only the multinational corporations. The corporations that sell expensive fertilizers to Third World farmers are the very same ones that buy up the low-cost wheat from the farmers to resell on the global wheat exchange.

The world is seeking a low-cost technology to reverse global warming. Why not consider photosynthesis and organic farming? Plants absorb billions upon billions of tons of atmospheric carbon dioxide, transforming it into plant biomass, food, and soil. Heritage wheat has at least 500 percent greater leaf surface area than modern wheat. Healthy soil itself is the largest known sponge for carbon dioxide, sequestering more than the atmosphere, plants, and trees combined.

GLUTEN INTOLERANT OR POISON INTOLERANT

Monsanto's Roundup weed poison, with its active ingredient glyphosate, is sprayed on conventional wheat crops to promote drying down of the plant prior to harvest. Could this be a contributing cause of the rise in wheat intolerance? Roundup not only destroys the beneficial bacteria in the human gut and contributes to permeability of the intestinal wall but causes autoimmune disease symptoms. Glyphosate residues are common in nonorganic wheat.

Glyphosate use has increased 400 percent in the past two decades. The rise in glyphosate use is equal to the rise in the amount of glyphosate found in sampled bread. Over a third of bread tested in 2013 in England contained measurable amounts of the weed killer.[6]

Never fear. Monsanto attempts to clear the air, explaining that "Roundup herbicides have a long history of safe use at home and in agricultural settings. The low levels of Roundup's glyphosate, which gives it weed-killing power, when ingested from the food we eat are well below what has been determined

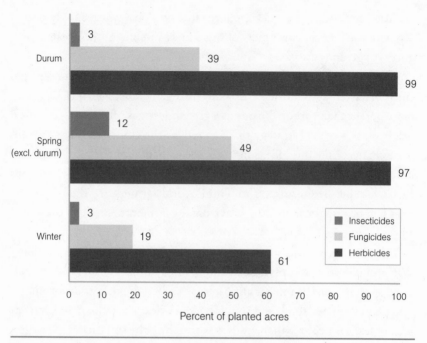

Pesticides applied to wheat planted acres, by type, 2012. *Source:* USDA, www.nass.usda .gov/Surveys/Guide_to_NASS_Surveys/Chemical_Use/2012_Wheat_Highlights/

acceptable for daily human consumption."[7] In the article "The Truth about Roundup and Wheat,"[8] Monsanto explains that their label's instructions for glyphosate advise to use it two weeks prior to harvest so that the poison has ample time to break down. Monsanto states that absolutely no poison can be absorbed into the wheat kernel. This comes as little reassurance.

Does Modern Wheat Make Us Fat?

In recent days, as the winter lies heavy in my corner of Western Massachusetts, I dance along with 1940s swing videos to keep moving indoors on frozen days. It is remarkable how thin and lithe folks were not so long ago. Why?

Green Revolution wheat has contributed to the global epidemic of obesity. Close to 70 percent of all Americans are overweight. Obesity worldwide has almost doubled to 34 percent in recent decades, according to studies conducted by researchers at Harvard University.[9] People may crave more calories because their food is less nutritious. Not only is the soil depleted, but modern grains' dwarfed root systems are unable to scavenge and process organic nutrients. Modern processed wheat spikes our blood sugar. Wheat gluten contains a unique carbohydrate called amylopectin-A, which sends blood sugar soaring higher than table sugar or a candy bar. Its high-glycemic carbohydrates stimulate our appetite with the druglike by-products of gluten proteins. We get fat.

What we have for breakfast influences how our body responds for the rest of the day. The type of toast you eat for breakfast affects how your body metabolizes food at lunch. Modern bread, even whole wheat bread, may be an *obesogen,* a food with a high glycemic index that raises your blood sugar level, then crashes it down, resulting in food cravings. We get food cravings despite our belly's being full because our system is out of balance from the blood sugar spike. Appetite is stimulated, not for more fish or chicken, but for quick carbs, more wheat, more candy, more soft drinks, more junk food. Impulse control is under the influence of a food craving resulting from an imbalance — after each indulgence.

ADULTERATION, CONVENIENCE, AND PROFIT

In the past two centuries not only has wheat undergone profound modification in its genetic makeup but the process of refining it into flour has borne equally significant changes. These alterations have changed the very nature of the bread we eat. If we want to embark on a revolution to transform the nature of the loaf, it will require two comprehensive transformations. The first, returning a wholesome process to the baking of bread, is well under way by artisan bakers. But the second, changing and restoring the nature of the grain itself, has barely begun. On one end of the continuum is my sourdough einkorn sprout bread, grain from traditional fields, fresh-milled daily, fermented, alive, handcrafted, and baked in a wood-burning oven. On the other end of the continuum is the Wonder Bread found on supermarket shelves.

In the 1839 book *The Good Housekeeper,*[10] Sarah Hale wrote:

> Fresh ground flour makes the best and sweetest bread. If you live near a mill, never have more than one or two bushels ground into flour at a time. Much damaged flour, sour, musty, or left in the field too long

in wet weather, so the grain swells and is ruined, is often used by the public bakers, particularly in scarce or bad seasons. The skill of the baker and the use of certain ingredients (alum, ammonia, sulphate of zinc, and even sulphate of copper have been used) will make this flour into light, white bread. It is nearly tasteless, and not as healthy or nutritious as bread made from the flour of good, sound wheat baked at home without any mixture of correctives.

Hale warned homemakers to watch for ground stones, bones, and plaster of Paris added to flour "to swell its bulk and weight and to whiten the color." To test the purity of flour Hale recommended adding drops of lemon juice or vinegar, which would cause flour that contains stone dust or plaster of Paris to effervesce. When a handful of pure flour is squeezed tight, it should keep the impression of the hand when the fist is opened, whereas adulterated flour will crumble.

Hard, indigestible bran was not a problem in wheat until the introduction of the modern combine placed convenience over quality. Farmers have known from the earliest days that the wheat plant should be harvested about two weeks before full maturity, when it is still slightly green at the top, then cured in stooks to dry down. Harvesting at full maturity results in a hard bran that needs to be sifted out, and a less nutritious white flour. Harvesting at the proper time gives a softer, digestible bran, richer flavor, and higher nutrition.

Before the Fleischmann brothers established the first commercial yeast factory in 1868, bakers either made their own sourdough starter or used "beer barm,"[11] the foam that rises on the top, or the sediments that fall to the bottom, when brewing beer. Historically, barm starter is from the first stage of making beer. Sprouted grain, known as malt, is gently simmered in warm water, then allowed to cool. The enzymes from malt's sprouting process are harnessed to make the barm that rises the bread. Hale recommended that if beer is "well brewed and kept in a clean cask, the settlings are the best yeast."

Sourdough fermentation is too time-consuming for industrial bakers. It takes only two minutes to produce air bubbles with ammonium chloride, a chemical used in the manufacture of engine antifreeze, and whipped with potassium bromate and potassium iodate to incorporate air bubbles.

Potassium bromate is a chemical used in permanent hair wave products. The more bubbles in a loaf of bread the less flour it contains and the more profitable for the baker.

Whole-grain flour is a product best enjoyed fresh-milled. The vital wheat germ and oils in flour will go rancid if left on the grocery shelves unrefrigerated. How could there be a centralized wheat system if flour needed to be fresh like a vegetable? In the late 1800s a milling method was invented to overcome the problem of how to prevent flour from going rancid so that it could be distributed nationally. A sifting roller mill system was designed that used grooved metal rolls to remove the germ and bran from grain. Roller mills separate out the inner endosperm of white flour, leaving a denuded, lifeless starch. The nutrient-rich wheat germ and outer bran layer were fed to animals. The refined endosperm-only white flour with a light, fluffy texture baked into bread whiter than people had ever seen. Roller mills rapidly replaced stone mills throughout the United States. To make the texture even lighter, a purified yeast isolate was combined with refined sugar in the dough. In the early 1900s, cheap fats made from hydrogenated vegetable oils were introduced to texturize the bread. Toxic chemicals were used to whiten and preserve the flour. Unless marketed as "unbleached," white flour is bleached with chlorine dioxide gas that forms dichlorostearic acid and methionine sulfoxide in the flour. It is known to cause nervous system damage in humans.[12] Supermarket white breads contain polyoxyethylene monostearate, a chemical that retains water so that the bread stays moist. This chemical causes cancer in laboratory rats. In Europe it is illegal to bleach flour.

White flour is vulnerable to fungal mold. Propionic acid, an antifungal ingredient used in athlete's foot powder, is added to prevent mold. "Enriched" white flour contains synthetic vitamins and minerals added after the bread has been stripped of natural vitamins and minerals. "Fiber added" refers to the addition of hard-to-digest vegetable fiber or synthetic methyl cellulose.[13] Read supermarket bread labels to see for yourself.

THE BREAD REVOLUTION: LIGHT-IMBUED GRAINS

The second aspect of the bread revolution, changing the nature of the grains themselves, is the heart of this book. How is it possible, in the context of the dramatic increase of gluten-related illnesses, that so few people have connected the industrial wheat plant with wheat-related allergies? Because few of us have ever seen a heritage wheat plant or tasted sourdough heritage wheat bread.

Humankind evolved with staple foods nourished by photosynthesis, the natural transformation of sunlight into nutrients. Heritage wheat has at least 500 percent more leaf surface area than modern wheat, reaching a majestic height averaging five feet. Greater surface area enables the photosynthetic activity that produces complex phytonutrients in the grains. Heritage wheat evolved over millennia to efficiently transform sunlight into nourishing food for humans through photosynthesis. Modern dwarf wheats lack sufficient leaves for the full processes of photosynthesis. They are less able to develop micronutrients and light-imbued proteins. Dwarfed modern wheats not only absorb less light for the photosynthesis that imparts fuller flavor, but their stunted roots have lost their evolutionary capacity to process organic soil nutrients. Humankind coevolved with heritage grain proteins. Human digestive enzymes have not evolved to keep pace with the high amounts of gluten in modern wheats. Sun-drenched, tall heritage wheats produce vital, light-imbued phytonutrients.

Biodynamic farmers believe that hardened modern grains contribute to hardened, materialistic thinking, whereas light-filled ancient grains nourish the human organism's etheric energy, promoting greater health, vitality, and well-being.

Try it yourself: Place einkorn grain under your tongue. It melts into a sustaining food. You can feel it. You can chew it without having to boil it first. It is a living food. Its baker was the sun. Its oven, the field. Place a grain of modern wheat under your tongue. It stays in your mouth, cold and hard as a stone. If you can chew it without breaking a tooth, you will get something that is more like bubblegum than food, since industrial breeders aim for high amounts of gluten, which makes stretchy bubbles for light, fluffy white bread. In fact, wheat breeders use the "chew test" in selecting plants to save for seed. The wheats that make the best bubblegum are selected. Modern baking quality is evaluated by loaf volume and gluten strength: bubbles and glue.

Nicolas Supiot, a *paysanne-boulanger,* an artisan farmer-baker in France, explains:

> In France, wheat is traditionally considered good for bread at an
> 11 percent gluten protein. It is hard to believe for me, but the gluten

strength of wheat today is approximately four times what it was 100 years ago! Industrial baking machines harshly beat dough requiring breeders to develop high gluten wheats to retain sufficient loaf volume after a destructive, intensive machine kneading.[14]

Heritage wheats' complex traits that enhance flavor and nutrition are the very qualities that have been bred out of the modern high-yield wheats. This is why renowned French artisan bakers such as Nicolas and Jean Francois Berthelot[15] prefer heritage wheats' superior flavor and baking qualities. Traditional artisan bakers ferment and knead by gently folding, then they let the dough rest over time to naturally knit together the delicate gluten strands.

WHAT IS GLUTEN?

Legend attributes the discovery of gluten to Buddhist monks who observed that when bread dough was immersed in water, the starchy flour washed away, leaving an elastic glob of gluten. This food, known as seitan, is a high-protein fare enjoyed through the centuries by monks.

Gluten is the protein-rich molecular matrix that transforms a mixture of flour and water into an elastic, responsive dough.[16] As dough rises due to the air bubbles generated by yeast microorganisms, gluten holds the dough together, giving bread its shape and fluffy, crumbly texture. Gluten consists of gliadin and glutenin. Gliadin and glutenin differ in the size and weight of their molecules. Modern bread wheat has been bred for a higher proportion of high-molecular-weight glutenin, which imparts elastic dough strength. Wheat landraces that have not been industrially bred have less glutenin, and therefore a more fragile gluten structure. It is not exactly the same as having less gluten. Most have very high gluten content. Einkorn's delicate gluten has a different quality from all other wheats because it is the only grain classified as a wheat that did not evolve from wild emmer wheat (*Triticum dicoccoides*).

How is gluten developed?

Physical kneading strengthens gluten adhesion by stretching and folding for a cross-linked, cohesive mass with greater bonding. The dough starts out lumpy and gradually bonds smoothly together with spring-back if poked. When gluten is well developed, the dough will stretch into a paper-thin film without breaking.

Chemical dough enhancers strengthen gluten adhesion with oxidants and reductants, as done in supermarket-scale bakeries. With the chemical enhancers, an intensive, short period of mixing develops the dough almost instantly! Modern white wheat flour is aged or brominated to oxidize the flour to help form gluten bonds and bleach out the yellow carotenoid pigments. Chemical dough enhancers also include natural conditioners like ascorbic acid (vitamin C) that facilitate the gluten reaction. Adding lecithin emulsifies the fats in the bread and retains moisture. Bean flour incorporates gluten-enhancing oxidants. Artisan bakers may add fava bean or soybean flour, which has lipoxygenase that oxidizes flour, to enhance gluten.

Biological fermentation develops gluten over time. A long fermentation period, called the "sponge stage," not only strengthens the gluten bonds but encourages microbes to digest the flour into a form with greater available nutrition and flavor. Slow fermentation combined with physical kneading and folding dramatically enhances the gluten adhesion, depth of flavor, nutrition, and texture of bread.

SKINNY BREAD: SOURDOUGH EINKORN BREAD SATISFIES LONGER

Einkorn sourdough breads combine the gluten-safe quality of this ancient grain with the enlivening action of fermentation to produce a flavorful, light bread with the lowest glycemic index of all. Einkorn has the safest, easy-to-digest gluten of all wheat species. Einkorn expresses the lowest celiac disease–causing components (known as epitopes) of all wheat species. After eight individuals with celiac disease consumed gliadin extracted from einkorn, their intestinal biopsies confirmed no negative reaction. In a rare *in vivo* study, twelve celiac patients experienced no difference in gastrointestinal complaints after twenty-eight days consuming either 2.5 grams of rice or einkorn daily.[17] That said, we advise anyone with gluten allergies to consult with your doctor to test if einkorn is safe for you. If you have celiac disease, beware; einkorn contains gluten.

Traditional sourdough einkorn bread is not only easier to digest but promotes a lower, steady-release blood sugar level that satisfies longer. The benefits of sourdough einkorn bread are highlighted in a study conducted by Danish scientists who tested if sourdough einkorn bread reduces the insulin and glucose responses compared to the responses to bread made

Let Us Look Deeper

What are the inner effects of redesigning humankind's staple food for industrial convenience? Why has gluten become toxic to more people today? Wheat bread itself has sustained robust traditional peoples from the Middle East to Europe for millennia. What is the problem now?

According to Dr. Karl-Josef Mueller, director of Biodynamic Cereal Breeding in Darzau, Germany,

> Gluten enables the absorption of wheat protein for human nutrition. If a chemical nitrogen-based protein is directly absorbed, the person will become ill. The protein is hardened and lacks a biological transformation. Protein must be transformed from a foreign substance that is unprocessed by biological activity into a human-available substance that is processed by beneficial bacteria. The metabolism starts with the bacteria in active plant roots and completes in the human digestive system bacteria. Gluten proteins have to be biologically broken down and then reconstructed inside the human body. It may never be possible to heal celiac disease, even with einkorn. But it may be possible to avoid getting an allergy in the first place with wheats better suited for human nutrition. We do not need hardening. We need an opening and widening of our senses, a kind of devotion in our feelings towards the world and its beings. When modern wheat and human inner needs are now in opposition, an incompatibility arises. The alarming rise in gluten allergies reflects this.[18]

with modern wheat. They documented that sourdough einkorn whole grain bread significantly *lowered* blood sugar levels compared to conventional yeasted modern bread that *raised* blood sugar levels.[19]

There are precious other forgotten wheat species that have yet to be restored that share the rich flavor, nutrition, robust growth habits, and gluten-safe qualities of einkorn. These are the very grains rejected by the modern industrial wheat system because they have not been bred with high levels of indigestible gluten for mass-produced breads.

Restoring Wheat Terroir

Terroir, a French word meaning *taste of the land*, expresses how the unique climate and soil of a farm imparts distinctive qualities to foods such as French wine, artisan Vermont cheese, or the rich flavor of coffee grown in its ancestral homeland of Ethiopia. Terroir is the dynamic interaction of the environment, the land, the seed's history, the farmer, cheesemaker, or brewer that give an artisan product its signature quality and taste.

Can mass-produced modern wheat bread from who-knows-where have terroir? It cannot. The baking industry requires uniform flour blends for mass-produced identical breads all in a row. The land is an endless field to the horizon. The farmers are invisible. Flavor is forgotten.

An understanding of terroir can give special meaning to the unique history, farmer, and "taste of the land." Is the soil fertilized by synthetic chemicals, or is it enriched with the vitality of biological relationships? The quality of the fertilizer is directly transferred to protein and the flavor in the grain. What is the historic tale told by the seed in the farmer's hands? The ultimate expression of grain terroir and flour quality is evoked by the quality of the mill and the unique baking style of an artisan baker. As we explore the biodiversity and heritage of landrace wheat, the potential to reinvigorate wheat terroir can come alive in our hands, farms, and bakeries and on our tables.

CHAPTER 2

Forgotten Grains

History celebrates the battlefields whereon we meet our death, but scorns to speak of the plowed fields whereby we thrive. It knows the names of kings' bastards but cannot tell us the origins of wheat. That is the way of human folly.

— JEAN HENRI FABRE

I n 1990, armed with a biodynamic training course in dryland farming and a certificate in wastewater treatment, I moved to the Middle East, to my ancestral homeland of Israel with the goal of finding partners for peace building through organic farming. I contacted a Palestinian farming organization recommended by John Jeavons of Ecology Action (growbiointensive.org). In the first week in Palestine, I was invited to the home of the Palestinian minister of agriculture and introduced to key farmers' organizations. There was great willingness to cooperate for the benefit of Palestinian farmers. Through intensive discussions, a core team developed a project. I wrote a proposal that was funded by the International Development Research Centre (idrc.ca), a Canadian sustainable development organization. Our goal was ecological water management for sustainable agriculture. I worked with American Near East Refugee Aid (ANERA) and other organizations to create field sites. There in the ancient village of Artas near Bethlehem, I saw for the first time the age-old arts of seed saving. My eyes opened wide as I watched traditional farmers collecting the healthiest plants to save for seed, rubbing the brassica seeds out of the pods into buckets, and fermenting the tomato seeds. Never in my hometown of Montpelier, Vermont — never in the United States — had I seen seed saving so seamlessly integrated into the farm system. Lettuce grows wild in the hills of Jerusalem and Bethlehem. Wheat also grows wild, as I was soon to discover.

Discovering Landraces

The difference between locally adapted landrace food crops and modern cultivars was dramatically revealed to me when I worked in organic extension in the West Bank. In scorching August, when the summer heat was at its zenith, I passed farm after farm of nonirrigated fields awash with verdant vegetables deeply rooted in the soil. How was this possible? In the cities the wilted summer gardens of modern cultivars barely survived the torrid heat even with irrigation, yet here in the windswept hills outside of Jerusalem, although the land was parched, the fields remained green. Before long I was in those fields with farmers exchanging seeds, sharing my open-pollinated heirloom seeds from Fedco Seeds (fedcoseeds.com) in exchange for their ancient landraces, and trading seed-saving tips.

WHEAT SHEAVES AND MATZAH TALES

Most commercial matzah in the United States is baked without human hands, a product of cold machines designed for industrial efficiency, destined for tables far removed from traditions of hearth-baked matzah. Only in Hasidic communities can one still find hand-baked hearth matzah. Until a hundred years ago all matzahs were round, hand-baked in wood-burning ovens in the village or home. The invention of the matzah machine by Dov Behr Manischewitz in 1888 changed everything. Rabbis of that day protested that matzah could not be kosher if it were mass-produced by a machine. Legend has it that Dov Behr started his matzah empire by selling matzah to Jewish pioneers traveling west in covered wagons as an easy food for the journey. The rest is history.

No machine-made matzah for me! As I baked matzah in my traditional wood-burning clay oven in Israel, carefully following the tradition, mixing deep and quick, rolling it out thin, baking it in the scorching hot oven, all in eighteen minutes, I kept wondering, kept searching, "Where is the ancient wheat for my matzah? Where can I find it?"

I searched everywhere to find local wheat for baking. After all, at the dawn of agriculture, the first wheats evolved in the Fertile Crescent. Search as I did, I was not able to find anywhere the emmer that sustained ancient Israel, that was used in the original matzah in ancient Egypt. I could not even find locally grown Israeli wheat for sale in Israel. Nothing. Israeli bakers nationwide explained that their wheat was imported from US mega-farms, with 15 percent of locally grown wheat mixed in. Even that 15 percent is a modern wheat bred for uniformity for the food industry. *Oy vey.*

Mount Tabor is mentioned in the Torah in Joshua 19:22, as the border of three tribes: Zebulun, Issachar and Naphtali. The mountain's importance stems from its strategic control of the junction of the Galilee's north-south route with the east-west road of the Jezreel Valley. In the days of the Second Temple, Mount Tabor was used to light beacons to inform the northern villages of Jewish holy days and of beginnings of each new lunar month.

I was hiking in the cool of morning at the foot of Mount Tabor. Glistening with morning dew as rays of dawn light seemed to illuminate the stalk with an inner glow, there it beckoned to me: an unusual plant at the field's edge, modest among the other wild plants. I plucked one stalk and carefully wrapped it in paper. I asked everyone about it, from the Arab farmers in the villages to Israeli researchers at the local university. No one could identify this unusual plant. It was beautiful to me. I kept it with me.

After fruitless searching for the elusive local wheat, I made an appointment with Dr. Rivka Hadas, director of the Israel Plant Gene Bank. I asked, were there any native Israeli wheats available today? Her response was troubling. Israel's native wheats are on the verge of extinction. Although the gene bank does store a small collection, the ancient varieties were collected decades ago; many have lost vitality, and vast populations of local, indigenous wheats have never been collected.

"Oh, by the way," I asked, as I pulled out my little seedy friend, carefully wrapped in cloth. "Can you tell me what this is?" "Where did you get this?" Dr. Hadas asked incredulously, her eyes wide with amazement. "It is *Em Ha'Hitah*, the wild mother of all cultivated wheat."

SEEDS WITHOUT BORDERS

I met Dr. Abdullah Jaradat at a Seeds and Breeds conference in Iowa in 2005. Abdullah, a senior USDA researcher, was instrumental in collecting ancient wheat for the Jordan gene bank. I contacted Abdullah. Could he help me find local wheat in the Middle East? Abdullah explained that Jordan and Palestine also buy 95 percent of their wheat from US mega-farms. Without hesitation, I e-mailed the Jordan gene bank, then phoned them. We spoke deeply. How could we cooperate to restore our shared ancient wheats? Their response was so warm that I invited them to meet in Jerusalem to discuss the problem. Incredible as it sounds, they accepted. The Israel gene bank was thrilled and helped raise funds to host them. Palestinian researchers agreed to join us. We were on!

Jordanian, Palestinian, and Israeli gene bank directors, traditional Arab farmers, and rabbis met together to discuss our loss of biodiversity. Indigenous seed knows no boundaries. We established a cooperative regional project, Restoring Ancient Wheat, to collect and exchange our remaining ancient wheats before they are lost to the world.[1] We need to rush. Time is short. The rare ancient wheats that remain are grown in remote villages. Families save a few shekels by buying the cheaper US wheat at their local markets, while their sons who grew up farming now look for work in the city.

Dr. Dominique Declaux, a French wheat breeder, joined us at the Restoring Ancient Wheat conference. I brought her to visit the Palestinian villages where I collected samples, and we discussed the critical need to develop markets as a strategy for genetic conservation. "Eat it to save it" became our motto. She suggested that I present the Restoring Ancient Wheat project at a European Union–wide wheat conservation and breeding conference being held the next week in Carcassonne, France, in 2005.

SOUND THE ALARM!

I was not working alone any longer. At the wheat meeting in La Besse, France, I met a lively group of young French farmers and bakers bemoaning the poor quality of bread made from modern flour and the loss of heritage wheat. Country after country reported on the sorry state of their wheat landraces. Dr. Kostas Koutis, director of Aegilops, a Greek biodiversity organization (aegilops.gr) told us, "Our landraces are lost, replaced by modern Green Revolution cultivars. Nothing is left on the farms." Dr. Anders Borgen of Denmark reported, "All our landrace wheats are gone. Nothing remains in the fields. We can only find landrace wheat in the gene bank." Georgia's Elkana Biological Farming Association lamented the tragic tale of how Russia occupied their land for long years, outlawing the cultivation of ancient Transcaucasus wheats. In every EU country, wheat landraces have been replaced by modern wheats.[2]

In Georgia alone, eight indigenous wheat species would have been lost to the world were it not for the work of Nikolai Vavilov, a seminal Russian seed saver who collected rare seeds during the early twentieth century. Until the 1960s, fourteen species of wheat, more than half of the major wheat species in the entire world, and 144 varieties were growing in the Caucasus region. These ancient wheats evolved in traditional organic farming systems that encouraged beneficial associations of wheat roots, mycorrhizal fungi, and other soil biota. Archaeological evidence suggests that agriculture evolved in this region at the same time that it did in the Fertile Crescent. Today this biodiversity has decreased to almost nothing.

In just fifty years, the irreplaceable traditional wheat populations have been lost to the world, save for tiny packets stored in the world's gene banks. The disappearance of traditional farming systems, the exodus of rural populations to the cities, and the industrialization of farming have caused the extinction of the world's wealth of wheat landraces. At least 75 percent of the genetic diversity of all crop plants was lost in the last century. At least 95 percent of the world wheat fields are covered with modern bread wheat, all made from the same species of wheat, *Triticum aestivum*, so ubiquitous that it is known as "common wheat." This was not always the case. Bread wheat became the predominant wheat only as a result of the expanding conquests of the Roman Empire, popularized by the Romans because its uniformity was well suited to the intensive large-scale farming practices that fed the vast Roman armies. Unlike the many forgotten species of wheat, bread wheat can be easily threshed to remove the grain from the husk. It is higher yielding and more efficient for harvesting and processing. However, it is less nutritious and less flavorful.

How Traditional Farmers Domesticate Plants

Domestication and adaptation are a dynamic process that is still happening to this day. We continue to domesticate wild foods today, just as traditional peoples have done. From the Middle East and North Africa to rural Europe, traditional people know and use local wild plants and animals. Domestication, cultivation, and breeding are integrated with gathering wild plants, seed saving, and hunting — all in the same farm by the same family. Domestication, bringing wild seed to the home garden and replanting it in the richer tilled soils, happens all the time. In my little farm in Israel, I constantly gathered wild foods, planted them in my fertile garden beds, made salads and soups from wild arugula, wild wheat, wild barley, wild chickpeas, purslane, nettle, wild mallow and more. Watching how easily the wild food adapted, self-seeded, and returned each year in abundance in my fertile garden soil, I understood the process of domestication as it unfolded before my eyes.

Today as I weed my garden in Colrain, Massachusetts, I bring in more "weeds" for dinner than any cultivated vegetable. They are tastier and thrive in the weather extremes when heirlooms wilt or drown. I leave the best for self-seeding. I form a soil circle around the weeds with the largest tender leaves and rich flavor, and do not pick them. These elite weeds are left to reseed themselves for the next season.

DID WHEAT DOMESTICATE PEOPLE?

The conventional understanding of anthropologists is that spanning 23,000 years ago in the ancient Fertile Crescent, 12,000 years ago in ancient Mesopotamia, and 8,000 years ago in Old Europe, people gathered wild wheats. They tasted and chewed. The seeds that scattered easily replanted naturally into the soil. Grains that clung more to the stalk were pulled off and brought home. Was it from a spilled gazelle-skin satchel or a tipped-over stone bowl? Did the children's play of digging into the soil cover the fallen seed with earth? After fall rains, did the moist grains sprout among the huts in front of wondering eyes?

As wheat was domesticated, this secure abundance sustained families, villages, and, soon, clusters of villages, as people needed to stay close to fields to tend and protect their grains from hungry animals and outsiders, soon nourishing the emergence of settlements. The domestication of wild emmer (*T. dicoccoides*), the mother of all cultivated wheats, and wild einkorn (*T. boeoticum*) was the catalyst for the birth of agriculture and the subsequent evolution of the biodiversity of landrace wheats.

Farming evolved as diversely as the many unique peoples living in ancient times. Many peoples worked in different ways to domesticate and grow diverse plants in various regions gradually over millennia. Some communities succeeded in growing their own food. Others failed. Those whose experiments flourished moved from a nomadic lifestyle to build permanent villages where civilization evolved. As nomadic people gradually settled and built more permanent homes to tend their grain, they began to keep animals in protective pens and evolved complex social systems. The ancient civilizations of Mesopotamia, Egypt, Israel, and Rome arose.

Do you know about Göbekli Tepe, the earliest known megalithic site, which is at least 12,000 years old? It predates the pyramids by over 6,000 years. Göbekli Tepe, located in modern-day Turkey, was built by a vast community of organized hunter-gatherers who worked together to create megalithic sacred temples over generations. Einkorn was domesticated, not by chance discovery but may be a result of organized food production by the builders of Göbekli Tepe. The need to feed the army of organized builders and worshipers may have resulted in the domestication of grain. "Conventional anthropologists thought that the domestication of wheat and civilization was driven by ecological forces such as the warming after the Ice Age. What we are learning in Göbekli Tepe is that civilization is a product of the human psyche itself, the striving for meaning that thereby created agriculture. This is the opposite of what anthropologists conventionally thought."[3]

What Is a Landrace?

Landrace seed is the living embodiment of the plant population's evolutionary and adaptive history, an ark of traits borne across generations into our hands. It is the expression of species interaction of the plant in its environment and the human culture that shapes it.

— Frank Morton[4]

Knowledge of crop plant origins is vitally important to preserve genetic diversity and to reinvigorate food crops that can weather the coming storm of climate change extremes. A landrace, also called a "farmer variety," is a population of plants or animals that evolved over generations of natural and human selection to be well adapted to its local environment. Landrace wheats are dynamic populations, constantly evolving, crossing naturally and responding to new challenges. Wide gene pools of landraces evolved survival mechanisms to produce stable yields and harbor broad genetic diversity and buffering capacity. As agriculture spread through Europe it was spread through landraces. As farmers migrated, their landrace populations came with them, adapting to new environments. Since the dawn of agriculture, over twenty species of landrace wheat evolved under natural and human selection. The evolutionary process is still underway, as changing pressures, climates, culinary cultures, and market criteria inform populations anew.

In the 1920s, the legendary Russian plant explorer Nikolai Vavilov identified global centers of crop diversity spanning Mesoamerica, the Mediterranean Basin, the Fertile Crescent, the Near East, highland Ethiopia, and China, where traditional farmers domesticated thousands of diverse heirloom folk varieties, or landraces. For example, native Andean farmers domesticated potatoes, beans, quinoa, and amaranth grains, as well as many tuber and leaf crops. India alone had at least 30,000 rice landraces earlier this century until the introduction of industrial monocropped systems.[5] The understanding of a dynamic, evolving landrace gene pool of diversity arose from observing how traditional farmers save and exchange seed in its environment.[6] Saving the seed alone is not enough. We need to restore the whole seed system of farmers, seeds, local markets, and traditional cuisine.

To traditional farmers and bakers throughout history, these wheat species, although rare to us, were as familiar as family. They were the daily bread. Only in this past century, as wheat was squeezed through a genetic bottleneck of uniformity for flour blends intended for mass-produced breads, has the wealth of wheat biodiversity been forgotten. In this light the opportunity to remember is a portal for each of us to restore our ancient family connections in a renewed coevolutionary relationship.

Landrace wheats are an integral part of traditional food cultures. Women are the seed savers in most traditional societies. In the Middle Eastern villages where I collected ancient wheats, each season at harvest time the grandmothers were brought out to the wheat field to select the best plants to save for seed. Culinary uses of landrace wheats by traditional peoples worldwide are based in the unique characteristics, flavors, textures, and colors of their heritage varieties. The modern emphasis of selecting food crops for yield and agronomic traits alone excludes essential aspects of producing food for nourishing, culturally diverse cuisines.

Compared to modern cultivars, wheat landraces have higher biomass; invest more root growth into deeper soil profiles; have increased ability to extract moisture from soil depths; have a far greater association with mycorrhizal fungi, which enables a vastly greater capacity to scavenge nutrients in lower-fertility soils; have higher transpiration efficiency to use more modest amounts of water, and have about 500 percent greater leaf surface area for photosynthesis, which produces easy-to-digest phytonutrients. Their increased concentration of soluble carbohydrates in the stem shortly after flowering ensures adequate translocation of nutrition to the developing grain. This helps combat late-season drought in arid regions. Under field conditions with limited nitrogen, wheat landraces absorb and translocate more nitrogen into the grain than modern varieties, due to greater preflowering nutrient uptake and an overall better buffering capacity to weather extremes. Tall height that towers above most weeds combines with allelopathic root exudates that suppress weeds: landraces have great weed-suppressive capacity. Landraces are a treasure trove of traits for higher nutrient and water uptake under the stresses of weather extremes.[7] I saw this for myself over the five years of organic trials in which modern wheat failed again and again, yet the landraces grew strong, producing stable yields under weather extremes.

Wheat's Family Tree

The wheat genus *Triticum* comprises at least twenty-five species in traditional classification according to shape and quality of the plant and seed. There are vast subspecies variations within each species. Botanists of the Greco-Roman period up to the nineteenth century divided wheats into two groups; *Triticum* encompassed the free-threshing wheats and *Zea,* the hulled ancient wheats. The modern classification is based on the discovery in the early twentieth century that wheat can be classified according to the number of sets of chromosomes in a cell. Wheat species sharing the same number of chromosomes are more closely related to each other.[8]

Overview of wheat species according to the number of chromosomes:*

Diploid (fourteen chromosomes)

Wild goatgrass (*Aegilops tauschii*), wild einkorn (*Triticum boeoticum, T. urartu*); cultivated: Einkorn (*T. monococcum* subsp. *monococcum, T. monococcum* subsp. *sinskajae*)

Tetraploid (twenty-eight chromosomes)

Triticum turgidum subspecies: wild emmer (*T. turgidum* subsp. *dicoccoides*) wild timopheevii (*T. timopheevii* subsp. *armeniacum*), cultivated: Emmer (*T. turgidum* subsp. *dicoccum*), Durum (*T. turgidum* subsp. *durum*), Persian or Dika[9] (*T. turgidum* subsp. *carthlicum*), Rivet or Cone (*T. turgidum* subsp. *polanicum*), Polish (*T. turgidum* subsp. *polonicum*), Khorasan a.k.a. Kamut (*T. turgidum* subsp. *turanicum*), Colchis Emmer (*T. turgidum* subsp. *palaeocolchicum*), Zanduri (*T. turgidum* subsp. *timopheevii, T. turgidum* subsp *militinae*)

Hexaploid (forty-two chromosomes)

Triticum aestivum subspecies: Bread wheat (*T. aestivum* subsp. *aestivum*), Club wheat (*T. aestivum* subsp. *compactum*), Indian or Shot (*T. aestivum* subsp. *sphaerococcum*), Macha (*T. aestivum* subsp. *macha*), Vavilovii (*T. aestivum* subsp. *vavilovii*) and Spelt (*T. aestivum* subsp. *spelta*), Zhukovskyi (*T. aestivum* subsp. *zhukovskyi*), Xinchan Rice (*T. aestivum* subsp. *petropavlovskyi*)

Following is a summary of major wheat species, but remember, this is a simplified overview of a dynamic, continually evolving, naturally crossing, vast expression of the *Triticum* tribe. Scores of diverse landrace wheat subspecies are not even mentioned.

Diploid (fourteen chromosomes)

Goatgrass (*Aegilops tauschii*). *Aegilops,* commonly known as goatgrass, is a wild diploid relative of *T. aestivum*. It naturally crossed with emmer to make bread wheat, contributing genes that give bread wheat its unique rising character.

Einkorn (*T. monococcum* subsp. *monococcum*). Einkorn is the most ancient wheat domesticated by hunter-gatherers and Neolithic farmers. It grows

---

* Blank common name indicates that no common name is in use in the English language.

wild in the area of the Karaca Dağ Mountains near Göbekli Tepe. Early Middle Eastern farmer-traders brought einkorn to Old Europe. Einkorn consists of the cultivated subspecies *T. monococcum* and the wild subspecies *aegilopoides* (*T. boeoticum*). *T. urartu* is a subspecies that grows in the Armenian Highlands, suggesting that ancient Caucasus farmers may have selected it into *T. timopheevii*.

Sinskajae, or Naked Einkorn (*T. monococcum* **subsp.** *sinskajae*). Sinskajae is an extremely rare, naked, free-threshing einkorn-type wheat discovered by Mikhailovich Zkukovskii on the coast of Turkey in 1926 and later found in Armenia during a collection expedition. It has compact ears, short awns, short stalks, low yield, and early maturity with looser hulls. It was more susceptible than einkorn to local diseases in my trials.

Tetraploid (twenty-eight chromosomes)

Em Ha'Hitah, Wild Emmer, also known as Wild Mother Wheat (*T. turgidum* **subsp.** *dicoccoides*). Tenacious "Wild Mother Wheat" (*T. dicoccoides*) evolved formidable resilience to harsh conditions over millennia of survival in tough climates and is the mother of all cultivated wheat. In the gradual process of domestication, modern wheat has lost traits that enable it to survive in the wild, such as storing protein and micronutrients in the seed. Wild emmer is more nutritious and higher in protein, zinc, and iron than any domesticated wheat.

Emmer was selected by ancient farmers for larger grains, evolving into the domesticated emmer. It was also selected for looser hulls that evolved into durum wheat. It is easy to see the relationship between wild emmer and the ancient Fertile Crescent *jaljuli* durum: an evolutionary link between wild emmer wheat and durum.

Emmer (*T. turgidum* **subsp.** *dicoccum*). Domesticated emmer, known as farro in Italy, and *shippon* in ancient Israel, was domesticated in the southern Fertile Crescent and cultivated in ancient times throughout the Middle East and Old Europe. It was the only wheat used in ancient Egypt and was used to make the first matzah. Emmer is gaining in popularity in the United States as a gourmet natural food. Emmer is higher in crude protein than bread wheat but has low dough extensibility. Emmer dough does not stretch and rise like bread wheat. The progenitor of durum, emmer makes excellent pasta and flatbreads.

Zanduri (*T. turgidum* subsp. *timopheevii*). Little-known zanduri (*T. timopheevii*) was domesticated from a wild grass (*T. araticum*) found in the Caucasus Mountains to Iran with its center of biodiversity in Georgia. Delicious, tall, and slender but with a fat seedhead, almost-wild zanduri, like einkorn, has exceptional resistance to diseases such as rust, fusarium, and pests. Since ancient times in Transcaucasian lands, bread baked with zanduri is known as the King's Bread because of its high quality and delicious flavor and is beloved as lavash flatbread. Although *T. timopheevii* is classified as a tetraploid, crosses of it with other tetraploid wheats are sterile. This suggests that it evolved in the Caucasus Mountains isolated from other tetraploids. We encourage seed savers to restore this forgotten delicious and resistant ancient wheat.

Mirabil (*T. turgidum* subsp. *mirabilis*). Mirabil, pharaoh, or miracle wheat is a branched poulard-type wheat known in England and France in the time of Philippe-Victoire de Vilmorin (1776–1862), a French seedsman who built the renowned Vilmorin seed company.

Poulard (*Triticum turgidum* subsp. *turgidum*). The poulards, closely related to emmer, have spectacular large seedheads with varieties of branched or composite shapes and generous, fat, round seeds. Poulards grow in the Mediterranean to middle Europe spanning the Balkans and Italy and stretching into France, Spain, England, Germany, Switzerland, Asia, and even Ethiopia. *Triticum turgidum,* known as rivet or cone wheat in England, where it evolved greater cold-hardiness than the Mediterranean poulards, first appeared in late Saxon times in England. Rivet was used in thatched roofs dating from the 1400s. Until the end of the Napoleonic Wars and into the mid-1800s, rivet was a predominant wheat in England. Although not used for bread making in the United States due to its lack of extensible stretchy flour for rising, it offers us huge untapped culinary potential.

Polish (*T. turgidum* subsp. *polonicum*). Large-grained Polish wheat thrives in warm climates and is especially grown in the southern Europe and Mediterranean region. Most varieties have a high protein content (27 percent) and are rich in gluten for bread. The large seeds are traditionally cooked as a cereal, ground into flour for noodles or flatbread, and have a rich flavor. It was rarely grown in Poland, despite its name – another grain mystery to solve! Beautiful and decorative.

Persian (*T. turgidum* subsp. *carthlicum*). Persian wheat, found by Nikolai Vavilov from Iran to the Caucasus mountains, is an ancient spring wheat species noted for its great immunity to diseases and delicious flavor. Known as dika in the Republic of Georgia, "it is likely that Persian wheat is a natural cross of bread wheat which it resembles with ancient emmer that contributes its great resistance to rust and mildew and full flavor."[10]

Next to bread wheat, dika ranks second in traditional Georgian agriculture, often grown in mixtures with bread wheat to protect them from disease and enhance their flavor and baking quality. It has a great diversity of shapes and colors ranging from red to white kernels. Although dika wheat was widely grown in Georgia for millennia, today it is critically endangered. The reason is the forcible migration of the mountain population to lowland urban regions, and the modernization of farming systems into monocropping. *T. carthlicum* is the most fusarium-resistant wheat in the world and urgently needs to be restored. Fusarium is the greatest wheat disease facing New England farmers.

Khorasan (*T. turgidum* subsp. *turanicum*). Khorasan closely resembles *T. polonicum* but lacks the long glumes. Research suggests that Khorasan may be a natural hybrid between durum and polonicum with large, long grains. It grows in abundance in Iran and is planted as a spring wheat in colder climates. Nutritionally superior to bread wheat, it has a rich, buttery flavor, is easy to digest, and is high in minerals. It contains more amino acids than bread wheat. Its protein level is higher than bread wheat. Because of its higher percentage of lipids, which produce more energy than carbohydrates, Khorasan is a high-energy grain. Khorasan is marketed in the United States under the trademarked name Kamut.

Durum (*T. turgidum* subsp. *durum*). Drought-hardy durum, whose name means "hard" in Latin, is the most widely cultivated wheat after bread wheat. Durum evolved from emmer that was selected for free-threshing hulls. Though high in protein, durum is low in gluten and is used for noodles, flatbreads, bulgur, couscous, tabouleh, and soups.

Colchis Emmer (*T. turgidum* subsp. *paleocolchicum*). Kolkhuri asli, or Colchian emmer, is a rare indigenous wheat of Georgia. It was discovered in archaeological excavations dating to the Neolithic Age: a living relic of prehistory. It is frequently found in mixtures with Georgian macha

29

wheat. Of the many favorable traits of kolkhuri asli, its resistance to fungal diseases is noteworthy. Its grain protein averages 18.8 percent and lysine 2.9 percent. Its high-quality gluten gives good bread-baking properties and flavor. The species has broad adaptability to the humid climate of west Georgia, also performing well under dry, hot climate conditions of east Georgia. This suggests a wide distribution in earlier times. The renowned Georgian archaeo-botanist V. Menabde considered kolkhuri asli as the ancient initial species that arose from the protomacha in the Neolithic, a lost treasure needing restoration.

Hexaploid (forty-two chromosomes)

Bread Wheat (*T. aestivum* subsp. *aestivum*). Bread wheat is the most widely grown wheat, covering over 95 percent of the world's wheat fields. Bread wheat evolved in Caucasus farmers' fields and is not found in the wild. It arises from a natural hybrid of emmer or durum with goatgrass (*Aegilops tauschii*). This versatile wheat has thousands of varieties and has adapted to diverse climates. It is sown in fall in mild lands such as France, and in spring in regions with harsh winters such as Scandinavia. It can be facultative, adapting to either winter or spring planting.

Hard Red Spring: Hard, high-protein wheat used for bread, high-gluten bagels, or pizza.

Hard Red Winter: Hard, high-protein wheat used for bread and to increase protein in pastry flour for piecrusts. All-purpose flours may be made from hard red winter wheat.

Soft Red Winter: Soft, low-protein wheat used for pastry.

Hard White: Hard, medium-protein wheat for bread and brewing.

Soft White: Soft, low-protein wheat used for pastry.

Club Wheat (*T. aestivum* subsp. *compactum*). Club wheat may be either of winter or spring growth habit. Stems vary in height but are generally stiff. Spikes are short, usually under 2½ inches in length, compact and flattened. Club wheat is a free-threshing wheat from primeval Europe, grown in Alpine countries, Austria, Switzerland, and southern Europe. It has a high protein content (22 percent) and is excellent for baking pastries. Its remains were discovered in archaeological excavations in the Georgian mountains dating back to the sixth millennium BCE. It was rarely found in pure sowings but grown in mixtures with bread wheat. It is grown today in the Pacific Northwest for pastry flour.

Macha (*T. aestivum* subsp. *macha*). Macha is a hulled bread-type wheat native to the Caucasus region, where it was found in great abundance by the renowned plant explorer, Nikolai Vavilov. He collected it in peasants' fields. Vavilov reported that the macha landrace had stable yields in extreme weather conditions and a high resistance to many diseases. V. Menabde suggested that the word *makha* refers to an ancient grain goddess and carried great totemic value to early farmers. It was commonly grown until the 1930s, after which it was replaced by higher-yielding modern wheat; needs restoration.

Spelt (*T. aestivum* subsp. *spelta*). Spelt, a hulled hexaploid wheat, has less gluten than bread wheat, making heavier breads with a rich, nutty flavor. The earliest evidence of spelt is from 6000 BCE in the Caucasus region and southeastern Europe. It gradually spread during the Bronze Age (4000–1000 BCE) throughout the Balkans and Europe. The wide distribution of spelt was facilitated by the migrations of early farmers. Eight hundred years ago, Hildegard de Bingen, a German mystic-botanist-artist, praised spelt's qualities:

> Spelt is the best of grains. It is rich, nourishing and milder than other grain. It produces a strong body and healthy blood to those who eat it and it makes the spirit light. If someone is ill, boil some spelt, mix it with egg and this will heal him like a fine ointment and make him cheerful.[11]

Spelt is a major cereal crop throughout southeastern Europe, primarily in Germany and Switzerland. Spelt holds its hull tight, making it more work to thresh.

Indian (*T. aestivum* subsp. *sphaerococcum*). Known as Indian or shot wheat, this rare species has been grown in India for millennia, was found in Neolithic settlements in Switzerland, but is almost lost today. Indian wheat is early flowering, resistant to yellow rust, and drought-tolerant, with higher protein than bread wheat. This round grain grows on a short stalk with fat seedheads. Archaeologists found *sphaerococcum* grains at Mohenjo-daro in the ancient Indus valley dating from 10,000 years ago. A similar round wheat, *T. parvicoccum,* was found in the ancient Israeli archaeological site of the village of Delilah, seductress of Samson, who fought the Philistines.

Vavilov's Wheat (*T. aestivum* subsp. *vavilovii*). Named after the famed Russian plant explorer Nikolai Vavilov, who identified this rare species, Vavilov's wheat has an irregular seedhead with slight branching. It is found in the Caucasus region.

Zhukovskyi (*T. aestivum* subsp. *zhukovskyi*). Zhukovskyi, a rare Georgian hulled grain may be a natural cross of *T. monococcum* × *T. timopheevii*, since it was discovered in the midst of a field of *T. timopheevii* and *T. monococcum* in 1926.

Xinchan Rice (*T. aestivum* subsp. *petropavlovskyi*). Rare Asian Xinchan Rice, despite its name, is actually a hexaploid wheat (*T. petropalovskyi*). It is a natural cross between polish wheat (*T. polonicum*) and bread wheat (*T. aestivum*). It has enormous kernels and glumes resembling Polish wheat but is hexaploid; it has better protein properties for baking. It is adapted to the dry land deserts of central Asia and needs to be selected in the moist New England climate for greater resistance to local diseases. Anders Borgen is selecting this rare species for adaptability to Scandinavia.

Rye and Barley

Although they are not in the wheat tribe, we would like to celebrate our friends, rye and barley.

Rye (*Secale cereale*). Rye is hardier than wheat and well adapted to northern climates such as Scandinavia. Although lacking gluten to rise, rye flour makes a dense, tasty bread. The grain is fed to livestock and used to make whiskey, gin, and kvass and is an excellent soil-building cover crop. The young, green plant is good forage for animals or dried for hay. Tough, dry rye straw is used for thatching; as bedding for animals; and in hats, mats, and paper.

Rye was known to the ancient Greeks and Romans but was not widely grown until the Middle Ages, when it grew as a weed intermixed with wheat in the same field and was harvested together as maslin, the wheat-rye mixed flour used for hearty peasant breads in Europe.

Two-Row Barley (*Hordeum vulgare*) and Six-Row Barley (*Hordeum vulgare*). The Hebrew word for barley, *seor*, meaning "hairy," well describes barley's generously endowed awns. Barley's great drought tolerance and ability to thrive in thin soils where wheat fails has made it a staple food of animals and peasants since the earliest farming. The mythic figure Gilgamesh drank barley beer and ate barley bread in ancient Mesopotamia. Around 11,000 years ago barley was cultivated in India. Barley was the staple of Plato and Aristotle in Greece. From biblical times to medieval Europe, barley was a staple food that was less costly than wheat. It is enjoyed in the dryland cuisines of nomadic Bedouins and Tibetans due to its drought tolerance. Barley is eaten in soup, bread, cakes, and the nomadic Tibetan staple *tsampa*, which is dried flour mixed with butter-tea. Malted barley ferments well for beer and distilled beverages. Barley water is still enjoyed today by the queen of England. Two- and six-row barleys are among the five ancient grains of Israel used for matzah.

LOST TREASURES OF COLCHIS: THE CAUCASUS CENTER OF WHEAT BIODIVERSITY

A Georgian legend recounts that when God was allocating land to the peoples of the world, he came upon the Georgians celebrating at a bountiful table. The Georgians were in a festive mood and invited the Creator to join in their drinking and singing. The Lord was so delighted that he gave these generous people the secret place on the earth reserved for himself in the Caucasus Mountains. Majestic Georgia is a land of contrasts, frozen mountains swept with wild gales and tumultuous rivers embracing warm fertile plains. The diverse geography has created a rich biodiversity with bountiful culinary traditions. In Georgian villages the living memory is vital, with chants and rituals honoring the weather spirits with vibrant polyphonic songs and dances dedicated to the sun. However, with the spread of media and globalization, old traditions are as endangered as seeds.[12]

When ancient Greeks and Phoenician maritime traders sailed the Mediterranean and Black Seas, the shores of Colchis (modern-day Georgia) were considered the end of their known world, Greek traders established three settlements on the Georgian coast of the Black Sea. Not only did they bring back tales of Jason, the Argonauts, and the Golden Fleece, but they also loaded their ships with the higher-yielding hull-less bread wheat of the Caucasus Mountains that was not yet known in Europe.

Fourteen wheat species have their ancestral homeland in the soil of Georgia; each generated many more localized landraces. Since ancient days, the Georgian landrace wheats evolved exemplary yield, flavors, and hardiness. Georgian agriculture is traced back to the sixth millennium BCE, when Kartvelian (East Georgian) tribes domesticated wheat, barley, oats, rye, and legumes such as peas, chickpeas, lentils, and fava beans. The rich diversity of fruit trees includes more than a hundred species of fruits, nuts, and berries.

Caucasus Hulled Wheat Species

T. monococcum: cultivated einkorn or gvatsa zanduri

T. dicoccum: cultivated emmer or asli

T. paleocolchicum: kolkhuri asli or Colchis emmer

T. timopheevii: chelta zanduri

T. macha: macha

T. spelta: spelt

T. zhukovskyi: Zhukovskyi

Causcasus Free-Threshing Wheat Species

T. aestivum: soft wheat (ipkli, khulugo, doli)

T. carthlicum: Persian (dika)

T. durum: durum (tavtukhi)

T. compactum: club (kondara, chagvera, nagala puri)

T. turgidum: English wheat

T. khorasan: Iranian hard wheat

T. polonicum: Polish wheat

Through the ages many empires have coveted the bounty of Georgia's majestic lands, which are rich with biodiversity. Her tenacious peasant farmers, known as the "farmers of the sword," developed a high art of vigorous sword dances within their harvest celebrations. Russia occupied Georgia since the beginning of the nineteenth century until 1991, except for a short four-year period of independence from 1917 to 1921. The long years of Russian domination struck at the heart of the Georgian people's deep love of their land by forbidding cultivation of landrace wheat. The Soviet iron hand sought to control her empire's people through their food system by eliminating local landraces from farmers' fields. Russia legislated that Georgia would be the fruit basket for the Soviet Union by exporting its luscious grapes, apples, peaches, and citrus. Landrace wheat production ceased. Only Green Revolution seeds were allowed to touch their soil. A handful of seeds of the world's most ancient landrace bread wheat were saved in the Georgian seed bank and Nikolai Vavilov's collection. Today, at the core of Georgia's recovery is the restoration of her heritage of diversified traditional agriculture and biodiversity.

Elkana, the Georgian Biological Farming Association, is working with their national gene bank to save Georgian landrace wheats and has restored enough to cover twenty-eight hectares with tsiteli doli landrace bread wheat (*T. aestivum*) and two hectares with dika (*T. carthlicum*).

In our SARE-funded Northeast Organic Wheat trials, Georgian wheats not only were totally free of fusarium, even in the rainiest seasons when other varieties were smitten, but had the rich flavor that artisan bakers search for. The Heritage Grain Conservancy is coordinating an Adopt-a-Crop program to restore and select these delicious, resistant landraces for our region.

IN-SITU OR EX-SITU CONSERVATION (IN THE FIELD OR IN THE FRIDGE)

In response to the critical loss of biodiversity over the past century, national seed banks were established worldwide. Although there are now more than a thousand seed banks, they maintain only a tiny portion of the world's food-crop biodiversity. Key regions have not yet collected their irreplaceable biodiversity. The problem is that landraces stored in gene banks become more uniform in the artificial conditions of cold storage and grow-outs in uniform, conventionally managed fields. Most concerning is that the millennia of evolution on farms in the hands of traditional farmers under natural weather stresses has ceased. Although pests and pathogens are rapidly evolving, the seeds stored in gene bank refrigerators are not. In addition, the collections of gene banks in eastern Europe and developing countries are chronically underfunded and poorly staffed and suffer power failures.

High-yielding, industrially bred seed has replaced farmer-saved seed the world over. Alarmed at the global loss of landrace seed, governments allocated funding to establish national gene banks, giant refrigerators to store landrace seed. This is known as *ex-situ* conservation, meaning "out of the natural environment" of the farmers' hands. Conserving landraces in gene banks is essential to preserve rare crops in a time when many farmers have forgotten how to save their own seed. The goal is to preserve varieties as they were collected, storing them as a frozen snapshot in time.[13] The US gene bank stores 59,880 accessions of wheat. Of those, 28,428 are landraces, yet only a handful of landrace wheats are commercially available in the United States.

Restoring Our Seed

We gardeners and farmers care about our direct relationship with soil, plants, and food. To grow plants from seed bought from others is one level of relationship. To grow plants from our own seed, to save seeds from our own plants, goes to a deeper level. It is fulfillment and continuity, plants and people maintaining each other, nurturing each other, evolving together. It is the unquenchable joy of doing what we are meant to do.[14]

— CAROL DEPPE

Seed saving, once an essential skill passed from generation to generation, is almost a lost art. Generations of farmers without advanced degrees not only produced their own seed but developed the food crops that we eat today. Seed catalogues are recent in the history of agriculture. The first seed companies in the United States were established in the late 1700s. A hundred years ago, when our great-grandparents were born, half of the people were farmers — one of every two people. Today only 2 percent are farmers. Where did all those farmers get their seed? Every farmer depended on the seed that they saved or was shared with neighbors. Although traditional farmers worldwide still grow, save, and improve their own seed, US farmers have abdicated their seed to "professionals." Public-university breeding programs, which introduced many locally adapted cultivars until late in the twentieth century, have almost disappeared, replaced by genetic research funded by corporations that patent the seed. A series of consolidations has rocked the seed industry, reducing the players to a small handful. One company, Seminis, controlled 60 percent of the North American seed market and was recently bought out by Monsanto. Multinational seed companies breed vegetables for uniformity, durability in shipping, and shelf life. Industrial wheat breeders select for uniformity, yield, and high gluten for fluffy breads like Wonder Bread. The large corporations who control the seed trade have bought out scores of small, regional seed companies. They cannot make enough money on the regional old varieties. The big bucks are in varieties with widespread adaptability. If we want varieties best adapted to our specific local conditions, we can get them only by working with our local seed networks and small seed companies, and by saving our own seed.

Agro-biodiversity is the foundation of food security, ecological health, and community vitality. In response to the loss of landrace seed, the on-farm seed conservation movement has arisen to restore landrace seeds within the dynamic farming systems where they evolved. Known as *in-situ* conservation, on-farm seed saving fosters a dynamic coevolution of landraces with local pests and pathogen complexes. Reinvigoration of landrace gene pools through on-farm conservation can be accomplished only in the conditions of the traditional fields and community systems where landrace crops evolved. Just as wild crops are genetic resources that cannot be contained in ex-situ facilities, ecological relationships such as gene flows between populations, natural adaptation to the environment, farmer selection, and culinary uses are integral components of a landrace crop's total evolutionary system. Restoring landrace wheat involves not only an understanding of population genetics but encompasses the social relationships that maintain healthy seed systems. Unlike ex-situ conservation of the plant outside its environment, farm-based conservation is a meta-disciplinary field involving indigenous knowledge, cultural practices, cuisine, and markets. It is about people and seeds.

When I worked with the Israeli, Palestinian, and Jordanian gene banks in the Restoring Ancient Wheat project, I distributed almost-extinct landrace wheat seeds to traditional farmers to trial on their farms. Most of these farmers had no access to irrigation. Their fields were watered by infrequent, quenching rains. Over the growing season, the deep-rooting old wheats reached down into the clay soil to the moisture in the lower levels. In those years there was a terrible drought. The wealthier farmers who could afford expensive fertilizers lost their crops of modern seed. One traditional farmer after another reported to me, "We lost everything except for the old wheats that you gave us. Those crops survived. That's when we realized how important they were for us."

Each season I collected samples of the new harvest from the farmers and trialed them at the gene bank fields to compare them to the original seeds that I gave the farmers. Year by year I documented the extraordinary increase in diversity and yield in the plots. The farmers' selections yielded a constant stream of renewed populations that adapted to the changing weather.

BREAKING THE MYTH OF THE FORGOTTEN CROP

Although more and more farmers understand the value of heirlooms and landraces, and are aware of the dangers of corporate control of the seed system, landrace wheat seed saving is a missing link. When I first started growing landrace wheats, extension agents said, "It is not economic to grow wheat in New England." I asked, "Economic for whom?" After four years of SARE-funded trials we documented that many landrace wheats are not only competitive with modern wheat in yield and disease resistance and have greater weed suppressive capacity but have far richer flavor, each with a fascinating history. We do not need waving fields of monocropped grain like the Midwest mega-farms.

Landrace wheat is a welcome addition to small-scale, diversified cropping rotations. It builds soil, breaks disease cycles of vegetable crop pests and pathogens, and yields enough for a year's supply of fresh-milled, homemade bread, or for a seed exchange or a backyard seed company — when using the methods taught in this book.

When I see the majestic fullness of heritage wheats given ample fertility and room to grow, I wonder in bewilderment how such beautiful plants could have been forgotten. Every farmer who is curious to grow heritage wheat can experiment in a small-scale garden where the soil is rich and well tended. Nothing I write can be as instructive in learning how to work with these old varieties as a few well-conducted experiments under your own hands and eyes.

CHALLENGE/OPPORTUNITY

Why are our vast treasures of delicious, disease-resistant, organic-adapted landrace wheats on the verge of extinction? Poulard and Polish wheats are closely related. *Triticum turanicum* subsp. Khorasan, a.k.a. Kamut, has a delicious nutty, buttery flavor and is a successful commercial grain today despite the USDA 1923 Wheat Bulletin #1340, which warns:

> This bulletin has been prepared to answer the frequent requests for information concerning the origin, productivity, and value of the varieties of Polish and Poulard wheats grown in this country and to warn farmers against paying high prices for seed of these nearly worthless grains. Poulard wheat usually produces low yields and is not suitable for making flour or semolina. Farmers are advised against buying and growing varieties of Polish and poulard wheat, as only unsatisfactory returns have been obtained in all parts of the United States. Man craves spectacular things even in a commonplace crop, such as wheat. Polish and poulard wheats are among our most spectacular cereal crops in appearance, and the stories which have accompanied the exploitation of these two grains would excite the interest of the most indifferent farmer. Neither of these wheats is of commercial value in America for commercial breads, but both have been offered to the buying public by unscrupulous or unknowing promoters who take advantage of their strikingly grand appearance.[15]

HOW TO EVOLVE NEW LANDRACES: RETURNING SEED TO THE HANDS OF FARMERS

On-farm breeding is a combination of art and science with the emphasis on ART. That means you rely on your intuition. You don't need to generate a table of numbers and run it through a statistical program to tell you which plant to take seed from. Will you get somewhere by relying on your intuition? Absolutely!

— DR. MARK HUTTON[16]

You do not need a PhD to restore landrace wheat. It takes intuition, fertile soil, and common sense. Traditional farmers often have an intuitive and profound knowledge of the whole plants on their farms as they observe them through the seasons. In combination with the thoughtful scientific approach of trialing, selecting, and documenting specific traits, combining natural selection and farmer selection we can develop new landraces

> If you're already running a market farm, you have the backbone to manage your own crop genetics. If you are a farmer surviving in this competitive and corporate era, you've got more than enough brain cells to manage your own soil fertility and your own crop genetics, thank you. We cannot afford to leave our seed development to the corporations anymore! Most importantly, seed-saving and crop improvement can be integrated readily into the seasonal operation of most market farms.[17]
>
> — Brett Grohsgal

adapted to our organic farms and markets. The value of landraces lies in their history of farmer selection and adaptation to diverse environmental conditions worldwide, their capacity to continually evolve, and their resilience to heterogeneous environmental conditions. The process of creating new landraces has largely been lost in industrialized agricultural systems.

Landrace wheats are an integral part of traditional food cultures. Culinary uses of landrace wheats by traditional peoples worldwide are based in the unique characteristics, flavors, textures, and colors of their heritage varieties. Selecting crops for yield and agronomic traits alone excludes essential aspects of producing food for nourishing, culturally diverse cuisines. Taste the wheat as you select it. Chew the kernels in the field to get a sense for the gluten strength and flavor hues. Speak to the plant, and listen to its response. Farmers are the original wheat breeders, inspecting their wheat plants day by day, tasting grain in the field, selecting plants for the traits best adapted to their farm conditions and artisan bakers' markets. In contrast to industrial breeding, farmers, seed savers, and artisan bakers can develop our own organic landrace wheat populations.

Wheat and all the food crops domesticated in the Fertile Crescent are self-pollinating plants. They do not require isolation distances. It is especially important to save seed of as large a population of plants as possible to maintain the population diversity. We can create "new" landraces by selecting, saving, and exchanging the seed, gene pools, and mixtures and saving seed of the diverse plants that thrive best in our fields. Increasing genetic diversity of wheat through the generation of multi-line gene pools or composite crosses[18] and use of landrace populations, in combination with introducing characteristics from modern wheat as appropriate, can be an effective strategy to increase yield in variable organic and low-input fields. Stable yields under low-input conditions tend to favor the polygenetic traits of landraces over modern pedigree varieties. Genetically diverse populations allow for adaptation through self-regulating, evolutionary systems that echo natural interactions that evolved landrace wheat characteristics, providing adaptable traits.[19] Biodiversity and genetic variability are the organic farmers' key defenses against disease and unanticipated stresses, such as weather extremes exacerbated by climate change. Flavor and baking traits, unlike yield and disease resistance, are not directly influenced by natural selection. Farmers need to work closely with bakers to select and save seed of varieties with delicious flavor and baking quality. Enhancing quality characteristics in a landrace population is essential for value-added organic markets. The raw material for selection is the genetic variation of natural adaptations, mutations, and composite-cross "new landraces."

RECURRENT MASS SELECTION

The age-old farmers' method of seed saving, known as recurrent mass selection,[20] involves saving a diversity of seed from outstanding plants in a field, bulking them together and multiplying the seeds. To evolve crops that can thrive in organic fields today, an on-farm breeding program using recurrent mass selection can increase yield, disease resistance, and quality.

The two basic approaches to selective seed saving in the field are:

Positive Selection: Save the seeds from the plants with traits that you like. This method may decrease diversity for future adaptability of the variety if insufficient numbers of plants are saved.

Negative Selection: Remove what you don't like. This approach enhances diversity in the population and is recommended to maintain landrace genetic complexity.

Today we have the opportunity to introduce greater diversity into our local food system by working with the vast treasures stored in gene banks.

Samples can be requested from all over the world, grown out, evaluated, and developed into new varieties adapted to our local climates.

GUIDELINES TO CROSS WHEAT IN A GREENHOUSE

Plant the seeds of the plants that you want to cross next to each other about 12 inches apart. Watch carefully as the plants mature, when you observe a stalk swelling with an emerging wheat spike. The "mother" spike is ready when the spikes are well formed. Clip off the awns on each spikelet. Oh so carefully pinch off the middle anther. Each spikelet has three florets with an anther. The delicate branching stigma reaches up seeking pollen. When the male anther sheds pollen and it is received by the stigma, the roundish creased ovule grows into a kernel. Gently slip around the leaflet on the two sides and pinch off the side anthers. Label a wax bag with the date. Cover the exposed mother spike with the wax bag. Seal with a paper clip. The pollen on the father spike is ready about a week later when a center anther pokes out of the spike. Select a father spike from a plant with traits you want. Snip, snip, snip off the awns. Watch the pollen-laden anthers rise up, seeking the female. Insert the exposed male into the bagged female. Shake in the pollen. Seal the bag and date. The hand-pollination process is complete.[21]

GUIDELINES TO CROSS WHEAT IN A FIELD

To cross two specific plants in the field, sow the two parents 12" apart, next to each other in your most fertile soil. This spacing should allow a head from the mother and a head from the father to fit into the same paper bag without cutting off the head of the father. In this way you will be almost independent of the time of flowering. You can emasculate a mother and choose any father that flowers later than the mother. The emasculated mother will wait for the father, and you will not need to watch and wait for the father's exact moment of pollen release. If they are not planted close to each other from the beginning, either dig up the entire father plant and replant it close to the mother and irrigate it to keep it alive until pollination is mature, or replant the father with the root system intact into a bucket with water, like a flower in a vase with soil.[22]

Composite-Cross Gene Pools
for New Landraces

A creative strategy to restore the rich diversity of landraces is to reintroduce new mixtures by combining hand-pollinated crosses of elite selected plants, then crossing with other elites in every possible combination to create a "new landrace." This is known as composite crossing. However, instead of selecting "promising" individual progeny to produce a pedigree uniform variety, the entire population is exposed to natural selection in the fields for subsequent generations. The new diversity is allowed to naturally evolve under the variable conditions of local organic fields. Composite-cross populations create dynamic gene pools that can reinvigorate landraces for continued evolution under the unpredictable stress conditions of climate change. For example, French farmer-bakers are working with French wheat breeders in an exciting participatory plant breeding program. Farmers and breeders confer together to identify desirable traits. The breeders cross the parents, then give the resulting gene pools to the farmers to select in their own fields. My Heritage Grain Conservancy "new landraces" program

The Creative Power of Selection

Today people in the industrial world are distant from both agriculture and nature. It is not surprising that few understand the power of selection. The raw material for selection is the natural genetic variation that evolved in landraces and that is created anew by mutations and adaptation. As selection is applied, plants with favorable traits are chosen. If the non-selected individuals are removed from the population, the remaining population will have a different gene frequency from the original population and selection will have been effective in improving the performance of the population. But, no new individuals or genotypes were created. What Darwin recognized and plant breeders harness is the creative power of selection.

— Dr. William Tracy, eminent plant breeder,
University of Wisconsin

How to Produce a New Variety of Wheat in 1868

Every variety of grain in cultivation will have ears that are totally unique, unlike any other in the field. The way to create a new variety of wheat is to go to a field of a good farmer with a well-known reputation for raising superior wheat. When the earliest of the grains are well ripened, select the choicest, whitest, plumpest heads to save for seed. Earliness of maturity is an important consideration when saving wheat seed. This will protect your crop in the future from the hardships borne by late rains. Procure a seed that is grown on high, dry ground, that is clean and white, so as to be free from moisture-afflicted diseases. You can select, if you choose, fine heads that appear to be quite unlike the greater proportion of the other heads. The ultimate product of these peculiar heads will be the new variety sought. Reject such heads as are not well filled out with plump kernels.

Prepare the ground with a thorough pulverizing and manuring. As the fertility of the soil becomes richer, the tendency to sporting increases in force, thus it is advised to maintain an abundant level of fertility, so that the inner qualities of the plant will emerge, to be considered in your selection process. Plant each seed one foot apart each way to the other, with one kernel in a place. Cover the seed with mellow, rich soil. If the ground is rich every kernel will tiller so profusely as to occupy the entire ground with large heads of grain. Next season and the two following seasons, weed the wheat and reject every head that appears a trifle smaller or different from the rest of the ears of your choice. In a few years the identity of the variety will be permanently established and the quality of the grain and its productiveness will be so greatly improved that one bushel of seed will yield several bushels more of superior grain per acre than can be grown on the same soil with ordinary seed. The pleasure, and in exceptional cases, the profit, to be derived is so considerable that the propagator of new varieties will generally be amply rewarded for the time devoted to this work.[23]

echoes this approach. Because organic farms have variable soils and complex environments, composite-cross gene pools with heterogeneous populations can best evolve localized adaptations in the hands of the farmers. Like a community bread oven that draws people into the joys of baking, a farmer-led breeding project can generate huge enthusiasm and local ownership and help restore a culture of generosity. Hold an abundance of seed in your hands and you will want to share it.[24]

RETURN TO RESISTANCE

Most of the plant breeding programs of the twentieth century have totally failed to achieve their objective of increasing resistance to disease and pests. We are actually increasing the susceptibility of many of our crops to disease and parasites.[25]

— DR. RAOUL ROBINSON

In his book *Return to Resistance*, Dr. Raoul Robinson explains how traditional seed saving by peasant farmers used "recurrent mass selection" to increase the plants' resistance to disease and pests. Traditional farmers grew mixed landrace populations with highly diverse gene pools. In a traditional field, the plants that are more susceptible to pests die before they can reproduce. In contrast, the modern use of herbicides and pesticides allows susceptible plants that would otherwise perish to reproduce and pass on their weak genes to the next generation. Pests and pathogens can rapidly mutate to overcome the resistance of modern uniform plants.

In his workshop on "Breeding for Durable Resistance"[26] in the SARE-funded Restoring Our Seed Conference of 2010, Dr. Robinson explained: "When I was an undergraduate, which was more than 60 years ago, I was taught that breeding plants for resistance to disease was a complete waste of time. Dr. Paul Muller in Switzerland had just been awarded the Nobel Prize for discovering DDT as an insecticide. We were taught to ignore breeding for resistance and to go with chemicals. Little has changed to this day.

"When I went to Kenya I fortunately had an experience which stood this on its head. In Kenya they started breeding wheat for resistance to wheat rust in 1926. In 1966 they found that the varieties they issued to farmers had an average commercial life of 4.5 years, then the resistance broke down. It takes eight years to breed a new wheat variety.

"We have two kinds of resistance: vertical resistance and horizontal resistance. Vertical resistance is controlled by single genes. It protects completely or not at all with no gradations. It is either present and active or not functioning at all. Horizontal resistance, on the other hand, is the dynamic interaction of a gene pool of diversity with a population of pests or pathogens, expressed with every degree of difference between a minimum and a maximum.

"Vertical resistance is single gene resistance using the gene-to-gene relationship. In nature each gene for resistance in the host is the equivalent of a tumbler in the lock. It would prevent an unauthorized key from turning that lock. In the parasite each gene is the equivalent of the notch in the key. It will enable the key to turn the lock. What happens when every door in the town has the same lock and every householder has the same key which fits every lock? The system of locking, that is, resistance, is ruined by uniformity; it ceases to exist. Yet that is exactly what we have done in agriculture by relying on vertical resistance. We've produced a new cultivar, but when every plant within that cultivar has the same lock, it is not very long before the parasite develops an entirely new population in which every individual has the same key, which matches that lock, and then you've got 100 percent matching, and that's the point where they say vertical resistance has broken down.

"Vertical resistance is unstable. It breaks down as new races of the pathogen evolve to survive the pesticide. This is a colossal disadvantage. Horizontal resistance is stable. That is critically important. It does not break down to new races or new strains of any parasite. This adds an important dimension to plant breeding because it means that if you get a good new cultivar with good horizontal resistance it need never be replaced. If it does get replaced it will only be with a cultivar that is superior to it probably in all respects. This means breeding for horizontal resistance is cumulative, progressing from poor to better to superior.

"Vertical resistance is expensive. It requires a team of highly trained scientists using costly laboratories and greenhouses. Horizontal resistance is cheap. If a group of us got together and wanted to breed wheat, we could do it and spend pennies. The investment is in the social aspect of team-building, farmer-to-farmer sharing of knowledge, and seed best adapted to local conditions. Any individual or group of seed savers can do it, possibly with some technical help from friendly people in the nearby university, but by and large can do it on their own.

"Vertical resistance is autocratic. If you are farmers in the third world, 95 percent of the wheats grown are industrial varieties bred with vertical resistance. The multinational seed companies are extremely proud of this. If I were a Third World scientist or farmer I would be extremely worried. It is like a dictator telling 95 percent of the people in the Third World what wheat varieties they will grow. This is terrible. When I first started publishing in 1970 I was regarded with extreme hostility. Now I'm getting wider acceptance but there are still some of the old diehards around who think that I am a dangerous lunatic."

CHAPTER 3

Landrace Grain Husbandry

L andraces are the quintessential crop for local organic food systems. Landrace grains represent the earliest stages of domestication cultivated in antiquity. They carry a veritable Noah's Ark of wild resiliences to adapt to a wide range of environmental challenges. Landraces evolved long before industrial breeding for global markets favored uniformity, appearance, and extended shelf life. The definition of *heirloom* as a variety that was grown over fifty years ago forgets the long history of landrace food crops. Heirloom food crops were selected from landraces, passed down from generation to generation to our hands.

A rich aspect of landraces is the stories they tell. Each seed bears a unique history of climate, soil, and village traditions of the local cuisine. The diversity of each field reflects the crop's variability and the plant's intelligent capacity to adapt to local conditions. Landraces were selected by farmers over generations to adapt to their local environment and culinary traditions and carried in migrations, trade, or conquests to new lands. Landrace complexity is not possible in a monocropped harvest streamlined for wide distribution. Unlike modern cultivars, with their uniform traits for industrial farms, landraces evolved for millennia in organic soils with deep roots for effective water and nutrient uptake. Their greater leaf surface area for photosynthesis generates abundant phytonutrients. Their deep roots scavenge for organic nutrients and water in the lower soil levels. Landraces' deeper connection to the soil, abundant leaves, and slow ripening contribute to their extraordinary flavor.

Because landrace grains were cultivated in antiquity before chemical soil amendments and pesticide spraying were introduced in the twentieth century, the question arises: How do we grow them today? What can we learn from the traditional grain growing methods? What tools and methods were used for weeding, watering, and harvesting that we can adapt to modern needs?

Landrace grain husbandry is an approach to increase the productivity and social value of wheat by enhancing the ecological dynamics within the soil, plant, and human systems. Our goal is to foster a culture of seed saving for robust wheat plants with greater adaptability to climate change. Landrace farming integrates ecological soil management and ecological plant breeding with community seed systems. A vital soil system nourishes larger roots. Deeper-rooted plants can reach lower soil moisture, a critical mechanism to avoid heat stress, and stabilize the plant, decreasing lodging (collapsing of the stalk) in rainy weather. With wider spacing and good soil tilth, landrace wheat roots grow deeper than in conventional dense spacing, enabling the plant to better survive the drought, heat, and rain extremes of climate change and produce high-yielding quality grains.

Nature has everything it needs in the soil of a wild ecosystem. It does not require outside fertilizers or insect and disease control by humans. So too, in a healthy farm ecosystem, as much as possible, fertilizers are produced and regenerated through cover cropping or composting on the farm without outside inputs. Our goal is to encourage a way of farming that fosters an intuitive, caring relationship with the whole farm — the soil, insects, crops, animals, environment, and community — so that all living organisms have a vital relationship with the crops.

However, in human-managed agro-ecosystems, much of the harvest is not recycled back to the soil but is sold as food products. Therefore we need to continually restore, amend, and build up the vitality, organic matter, and minerals in our soils that may have been depleted. Yet our goal is still to echo the natural interactions of the wild garden. A whole natural farming approach to wild habitats for pollinators, cover cropping, and natural cool decomposition processes is our model.

Weather and Wheat:
Adapting to Climate Change

Understanding that comprehensive changes are required to alleviate the root causes of global climate change, coping with today's unprecedented global warming requires adaptive strategies to maintain a secure local food supply. In the past forty years in New England, the weather has become warmer and drier, with more unpredictable, damaging heavy rain. The growing season has increased by three weeks to one month.[1] Our region is expected to experience more intense droughts punctuated by extreme weather events.

Our goal is to develop wheat mixtures with the buffering capacity to adapt and thrive in unprecedented weather extremes, to increase holistic cropping system diversity to mitigate climate change impacts, complemented with whole-farm permaculture-type landscaping that protects soil and channels rains for use in drought periods. Modern wheat is particularly vulnerable due to its pedigree uniformity and dwarfed roots. Landrace wheats have evolved adaptive survival traits over millennia, which produce stable yields under the weather extremes of climate change.

Integrating Grains in a Whole Farm System

Learning to grow landrace grains successfully is like learning to grow other crops in an organic farming system. The goal of an organic farmer is to create an environment where the intended crop is the best-adapted species for that location at that time. Whole-systems farming creates an environment that is uniquely suited to the crop being grown. Think long-term with the big picture.

ASKING THE RIGHT QUESTION

The process of learning starts with how the question is framed. Rather than asking, "How do I grow landrace grain?" ask instead, "How do I create a sustainable farming system where landrace grains thrive?" It doesn't work to think about how to grow any one species by itself. All of the diverse species in the production system are equally important. Construct the cropping system in such a way that each species is grown after its most suitable preceding crop.

51

THE ILLNESS SUGGESTS THE CURE

Farmers today are accustomed to look at weeds, pests, and diseases as random problems from which the crop needs to be rescued with pesticides or crop protectants. Likewise, farmers fill fertility requirements of crops with purchased inputs. Instead of focusing on the short-term challenges of how to deal with crop pests and fertility requirements, ask, "Why are these pests and deficiencies occurring?" What environment attracts the pests? What factors in the past management of the field created the conditions that encouraged the pests? Approach agronomic problems proactively rather than reactively by looking upstream to understand the causes of pests and develop holistic strategies for managing a farm. Cultural practices, crop sequence, soil management, and the past history of a field all combine to determine how well a crop will grow and what kinds of pest and disease challenges are likely. A change in approach is required. We need to replace a high-input, short-term, reactive management with a whole-farm systems approach to grow landrace grain species successfully.

Landrace Grain Husbandry Combines Three Dynamic Aspects:

Organic Soil Fertility: integrating cover cropping, intercropping, manure, compost, and generous mineral inputs in reduced-tillage, three-year rotations with a fourth year fallow

Seed Saving: nourishing the whole plant to reach its full potential, selecting and evolving genetically diverse landrace populations spaced wide in living soil systems

Community Seed Systems: restoring diverse landrace populations and seed exchanges to promote knowledge and local pride; fostering a unique "terroir" of the seed's history, "taste of the land," and farmer, community ovens, and educational programs in schools

Ecological Cropping Systems

As a society we have traded excellence for uniformity. When uniformity becomes the goal you lose the chance to be really great.

— *KLAAS MARTENS*

Manure, cover crops, compost, and mineral amendments provide fertility in organic production systems. Wheat thrives best in rotations of (1) cover crop/intercrop, (2) wheat, and (3) vegetables, enhanced by intercropping with clover and legumes. Building soil organic matter not only enhances nutrient availability but improves water-holding capacity, increases the cation-exchange capacity, and prevents soilborne diseases.

BALANCING NITROGEN

Enhancing the life processes that regulate the cycling of nutrients by cover cropping and building soil organic matter provides the balanced soil for wheat to thrive. Wheat prefers ample nitrogen, like corn, but the quality of the nitrogen makes all the difference. Adding too much nitrogen may cause lodging, increase disease, and decrease water use efficiency, resulting in yield reduction in dry seasons. Excessive nitrogen causes susceptibility to cold injury and to mites. Too much nitrogen promotes excessive vegetative growth that weakens kernel development, resulting in soft grains of low quality. Too little nitrogen not only reduces grain quality but weakens plant vigor, causing poor grain set, disease, or insect susceptibility that contributes to low yield. In cereals, fusarium and powdery mildew—severe diseases in New England's cool, moist climate—are encouraged by excessive nitrogen.[2] Wheat prefers firm clay loam soil with good drainage that holds the nutrients well. Wheat grown in sandy soil, although well drained, needs significant compost for ample nutrients.

Dr. William Albrecht described complex interrelationships between the availability of different minerals in the soil. He observed that having a high calcium level allowed plants to take in nitrogen in bigger quantities. Very high nitrogen levels, on the other hand, inhibited the uptake of many other minerals, including copper, iron, potassium, phosphorus, and manganese. Low copper levels are associated with increased pressure from fungal diseases and weak stems.[3]

Heritage wheats need less nitrogen than modern wheat. Farmers have long observed that putting on too much nitrogen causes weak stalks and lodging. Heirloom grains evolved large root systems that reach out to scavenge organic nutrients. They are not dependent on synthetic inputs for their fertility requirements. If you sow wheat in the fall, plan the season before to plant cover crops and add manure or compost and minerals that will release slowly over the coming season.

WINTER, SPRING, OR FACULTATIVE

A time to sow and a time to reap.
— ECCLESIASTES 3:2

In wheat's ancestral homeland in the Fertile Crescent, all seeds are planted in fall and harvested in early summer. Wheat later migrated northward and adapted as a spring crop to survive the harsh cold climates of Northern Europe, but its physiology is best suited to the fall-planted cycle.

Small grains can be categorized as winter, spring, or facultative. Most landrace grains are facultative, meaning that they can be successfully sown either in the fall or in the spring. Winter grains have a vernalization requirement for a minimum period of below-freezing weather after emergence before they can form heads and set seed. The vernalization requirement is adaptive for winter grains, preventing them from advancing too far in maturity before winter begins. If winter grains didn't vernalize, they could begin to form heads in the fall before winter if the weather was warm and the seeding date early. Vernalization also fosters winter hardiness. Winter grains have higher yield potential than spring types and suffer less competition from weeds but may have richer flavor and lower protein content. Seeding winter grains in the spring may result in a low-growing stand of grain that fails to head or produce grain.

Not only do winter wheats develop more extensive root systems for stable nutrient uptake, but their well-established root systems enable rapid spring growth that outcompetes most weeds. Winter wheats perform better than spring types in mid–New England. All of the wheats grown in colonial Massachusetts, for instance, were winter types from France and England, where primarily winter wheat is cultivated. Annual and perennial weeds are controlled with tillage before wheat is planted, and once it has become established, winter wheat can outcompete weeds.

Spring small grains are generally sown as early in the spring as weather and growing conditions allow. Spring varieties are adapted in regions where winter temperatures are too harsh for reliable winter survival.

Facultative grains have no vernalization requirement and therefore must be seeded late enough to prevent them from growing too big in the fall to avoid winter damage. A compromise between spring and fall types, they are generally best adapted to the warmer range of regions where winter grains can be grown. Facultative grains are higher in protein than winter grains and must be sown later than most winter grains.

PLANTING TIME FOR WINTER WHEAT

Early September is the optimal wheat planting period in southern to mid–New England. In northern Maine and Vermont late August is timely. Plant winter wheat early enough for the development of three to four leaves before winter, about a month before the first frost. Depending on the locality, this can range between September 1 and October 20 or the time of first frost. Earlier

Table 3.1. Three-Year Cover Crop, Wheat, Vegetable Rotations

Winter Wheat				
Year 1: Spring to Fall	Year 2: Early Spring	Year 2: Midsummer	Year 2: Late Summer	Year 3
Spring: Incorporate compost, amendments. Early Summer: Cover crop with legumes and mustard. Till in mid-August. Plant grains early Sept., then clover.	If you did not plant clover in the fall, seed clover as soon as snow is thawed in the wheat beds.	Walk your fields, saving seed from the healthiest, disease-free plants.	Plant late vegetables and/or fall cover crop.	Rotate different families of vege-tables. Intercrop legumes. Year 4: repeat year 1.
Facultative, Double Harvest,* Low Till				
Year 1: Spring	Year 1: Summer	Year 2: Summer	Year 3	
Plant wheat in early spring. Cover crop under wheat late spring.	Harvest wheat by scything close to ground. Leave roots in soil to winter over.	Roots regrow new strong seedheads. Harvest midsummer. Till in roots. Cover crop.	Till under cover crop and plant vegetables. Year 4: Cover crop and plant winter wheat, or plant faculta-tive in year 5.	

* The double harvest method is contributed by Nigel Tudor of Weatherbury Farm, Avella, PA.

planting will cause more disease problems, weaken plants that are prone to winterkill, and lower protein in the kernel due to overproduction of tillers.

Planting According to the Moon

Early farmers observed that each moon phase imparts an influence on the way plants grow. The moon cycles affect the rising and falling of the moisture in the ground and in the plant system. Planting by the phases of the moon harnesses the moon's gravitational influence on groundwater in the soil and moonlight on the leaves to improve the health and quality of plants.

Waxing is the phase when the moon visibly increases between the new and the full moon. When the moon is waxing, it is in its first and second quarters. In the *waning* phase between the full and the new moon the lunar light decreases. The waning of the moon occurs in the third and fourth quarters. The rising and falling of the soil moisture "tide" affects harvest and storage. If the crop is harvested at the optimal time of the lunar cycle, it will last longer. The rising of the soil tide influences how the plant stores water at different times of the lunar cycle.

As the waxing moon's light increases, it is an auspicious time to plant leafy crops, cereals, grains, and flowers that produce growth above the ground. Plant grains from the first to second quarter. Biodynamic researchers have observed that growth and liquid absorption peak at the full moon and drastically decline during the new moon. Traditionally, gardeners sow seeds right before and up to the full moon. Our ancestors observed that seeds germinate more rapidly right before the full moon. Some people today sow at the new moon in order to ensure germination before the plant's growth spurt caused by the full moon. Fruits or vegetables meant to be eaten immediately are at their best when gathered during the waxing moon. When the moon is full, plants are at their peak.

The waning moon is the best time to harvest grains. Herbs, especially medicinal herbs, will be more potent if picked at this time. Over centuries, farmers found that crops store better if harvested during the waning moon, when water content is decreased.

Root Systems

We seldom give enough thought to the unseen organisms around the roots and in the soil. We do see the crop plants, cover crop plants, and the weeds.

By observing our crops, we can see what the soil organisms do. Every plant has a unique community of organisms that live in and interact around its root system. The biologically diverse zone immediately around the root is called the rhizosphere. The organisms in the rhizosphere perform important functions that benefit the plants. Each plant species has its own unique community of rhizosphere organisms that benefit from the exudates given off by the roots of that particular plant. Up to half of the energy captured by a plant during photosynthesis is excreted in the form of root exudates that feed the rhizosphere organisms. The complex interactions between the millions of species that inhabit healthy soil in the root zone can be harnessed by the organic farmer. The beneficial organisms in the root zone of a healthy plant protect it from diseases. Others help the plant obtain minerals. Still others help it compete for resources over other plants.

The soil is changed by the action of plants and their associated species, building a more favorable environment for some species and less favorable for others over time. This process, together with the farmer's actions, such as adding soil amendments, doing tillage, harvesting, or removing crops, and seeding new crops, determines which species will be best suited to the soil. A sound crop rotation is dynamic, adapting to evolving soil conditions.

COVER CROPPING

Each crop should be grown after its most suitable predecessor.
— KLAAS MARTENS

How can we create a soil environment that best nourishes grains? Grains will thrive when they have a biological advantage over potential competitors. Modern winter wheats are top-dressed with nitrogen in the spring. In contrast, landrace winter wheats benefit little from top-dressing since their roots draw from deep in the soil.

Fertility management for landrace wheat starts the season before by incorporating manure, compost, and minerals in the cover crops to prepare the soil, since the landrace wheat roots grow significantly deeper. Landraces evolved aggressive nutrient-scavenging mechanisms over millennia in low-input organic soils. Even if there is an open field inviting you to plant wheat, don't. Wait. Take the extra time to prepare your soil for grain success.

First build the soil with cover crops, especially brassicas such as mustards, then plow under. Only after a season of cover cropping will the landrace grain thrive in the second season. The wait allows minerals to break down and be enlivened by biological activity, making them more available to the wheat. The complex soil ecology nourished by tilled-in cover crops provides slow-release living nutrients for winter wheats.

Cover cropping, a.k.a. green manuring, is the practice of growing and plowing in plants to feed the soil that feeds the plants. A successful cropping system harnesses the natural succession of species in the soil to provide a favorable environment for each crop in the rotation. The soil doesn't distinguish between crops and cover crops, or even weeds and crop plants. It is home to all of them, and whichever species is best suited to the conditions will grow best.

Cover crops reduce fertilizer costs by adding plant-based nitrogen and minerals, and build soil health by increasing soil organic matter. This enhances water holding and infiltration capacity. Many cover crops are allelopathic, exuding root chemicals that suppress competitive weeds.

Cover crops include legume plants such as clover, hairy vetch, peas, and alfalfa, which convert nitrogen in the air to a form that plants can absorb; and nonlegumes such as buckwheat, cereals, and crucifers like radish, turnip, and mustard, which add organic matter. Mixtures of both types of cover crops, such as peas with oats, winter peas with triticale, and oats with radish, are especially beneficial. When planning a rotation with wheat, it is helpful to divide cover crop species into two groups: (1) those that winter-kill, leaving ground-covering residue over the winter, and will not regrow in the spring, and (2) those that grow in the fall, go dormant during the winter, and then grow up again in the spring.

A legume cover crop adds significant nitrogen to the soil and can provide a major portion of the nitrogen required for landrace wheat.

Wheat prefers well-drained, medium- to heavy-textured soils with a pH in the 6.5–7 range. Winter wheat requires higher fertility than spring wheat. Winter wheat is a heavy feeder, although not as demanding as corn. It usually follows clover or soybeans in a rotation. Being a heavy feeder, winter wheat should not follow any crop like corn that also demands high fertility. In the fall, three weeks after sowing winter wheat, or in early spring, under-sow frost-planted clover or mustard into the wheat field. (See the "Intercropping" section on page 61.)

Winter-Kill Cover Crops

Deep-rooting *daikon radish* is a nontraditional cover crop that reduces soil compaction by opening up the lower layers, improving water infiltration, suppressing weeds and nematodes, and controlling erosion. Radishes bring up soil nutrients to the surface and increase soil organic matter by up to five tons per acre. Plant in the fall a month or two before the first killing frost. Tillage radish will grow rapidly through the fall, then winter-kill, leaving the soil in great shape for the following spring's cover crop.

Oats produce vigorous growth through the fall without making grain. Oats' natural allelopathy suppresses weeds while producing abundant biomass. During the fall, they sequester nitrogen, potassium, and phosphorus, holding it in a stable form through the winter. They decompose through the winter, leaving the soil mellow and ready to plant in spring. Oats can be grown alone or in combination with peas, turnips, or radish. Fall oats should be planted by the end of August, or forty to sixty days before the first killing frost.

Buckwheat is a quick-growing cover crop that suppresses weeds, brings up phosphorus, and improves soil tilth. Buckwheat is a great pioneer crop to restore neglected, weedy fields and helps prepare them for heavier-feeding wheat. Although usually planted in the summer, it can be successful as an early fall cover but will frost-kill.

Field peas grow vigorous vines that break down rapidly, releasing accumulated nitrogen in a form that other plants can easily use. Field peas, when mixed with oats in the fall, produce a thick, lush, nitrogenous biomass that builds soil fertility and tilth.

Yellow mustard helps control soilborne diseases by producing glucosinolates, hot-flavored chemicals that suppress fusarium, root rot, and nematodes in the soil. Rotating mustards before wheat and incorporating mustard seed meal helps reduce fusarium.[4] Although cover crops may not directly suppress fusarium, their key role in enhancing soil microbial communities contributes significantly to fusarium-suppressive soil that is antagonistic to the survival of fusarium (*F. graminearum*) in wheat residue.[5]

Overwintering Cover Crops

Rye, the hardiest of the cereals, produces great leaf biomass and extensive roots with strong allelopathic weed suppression, helps control nematodes, and builds up poor soil. Plow under in early spring before it grows too high.

Hairy vetch. Although vegetable farmers appreciate vetch for nitrogen fixation, soil building, and weed suppression, hairy vetch is a noxious weed in a grain field. Its life cycle matches that of winter wheat, producing round black seeds that are difficult to separate from wheat using normal cleaning equipment. Flour from wheat contaminated with vetch seed is bitter, with unattractive black specks. Vetch produces 10 to 20 percent hard seed each year; thus it is difficult to remove once established. Avoid hairy vetch in a grain rotation.

Triticale, a cross between wheat and rye, combines the vigor of rye with the growth habit of wheat. Triticale can easily be combined with peas for a cover crop mixture.

Austrian winter peas. Planted in September with barley or triticale, it grows slowly in fall, then explodes with vigorous growth in spring, fixing abundant nitrogen. A small grain is needed for support — without it, winter peas will quickly go down at bloom, as they become heavy with leaves and pods. Small grains help protect against winterkill. Klaas Martens recommends two bushels of barley per acre as about the right amount to provide adequate support for the peas without being too competitive.

Red clover is a dependable legume cover crop that farmers rely on for nitrogen and weed suppression. Clover can be frost-seeded into winter grains in early March. It grows through the spring, then covers the field after the grain harvest. By the following year, it can be plowed under, providing enough soil nitrogen to grow a healthy crop of corn or vegetables. Clover can also be sown with a spring small grain before grain emergence. Red clover can be drilled into no-till wheat stubble in the summer, or overseeded into soybeans in the early fall at leaf-yellowing. Soil diseases hosted by beans, peas, and soybeans are also hosted by clover, so avoid close rotation of these species. In the wide spacing of our system, we intercrop clover or another low-growing legume under the wheat to suppress weeds and build soil. Plant in the fall three weeks after sowing winter wheat, after the wheat seedlings are established, or in early spring under-sow frost-planted clover or mustard into the wheat field.

Hulled wheats are well suited to being grown after soybeans. Soybeans are often harvested too late to allow wheat to be planted in time. Hulled wheats are a better choice after the optimum time for wheat has passed. Winter grains should be frost-seeded with clover or other legumes in mid- to late winter. Delay seeding of alfalfa until there is no risk of frost. Grain that will be followed with a hay crop should also have grass in the hay. Grasses start faster if they are seeded along with winter grains in the fall. Legumes like clover and alfalfa should be frost-seeded in late winter. After wheat is harvested, the straw should be removed or chopped up finely.

INTERCROPPING

Intercropping grains and legumes harnesses ecological principles of diversity, species interaction, and natural biological controls to foster beneficial soil-plant interactions, resulting in greater yield stability than monocropping. Intercropping wheat with legumes draws on wider pools of nitrogen and phosphorus, enhancing niche complementarity. Synergies can be intensified by planting complex plant communities within livestock rotations, mimicking the ecological diversity of niches in natural systems.

Winter wheat can be overseeded into maturing soybeans. Plant when the soybean is at the "yellow leaf" stage, when 50 percent of its leaves have turned color from a month before harvest. Overseeding has the advantage of providing continuous ground cover and holding the nitrogen in the soil that was accumulated by the soybean plant.

Wheat grows poorly in land that has grown up in woody weeds and goldenrod. It also does not grow well in fields that have old grassy sod. Land that has old sod or woody species growing on it is beginning to revert to forest. The biology in such soils is not well suited for wheat.

The Nitrogen Cycle's Effect on Legumes

Each species takes specific nutrients from the soil and gives other things back. Each species changes the soil, making it better suited for different species. For example, legumes fix nitrogen and greatly increase the amount of nitrogen compounds in the soil. But legumes fix nitrogen inefficiently in soils that are already high in nitrogen. They can break the bonds between phosphates and calcium efficiently, making phosphorus that has been tied up in basic (high pH) conditions soluble. Pathogenic nematodes and fungi rapidly increase in numbers and begin to attack the roots of legumes after a short time. The high soil nitrogen levels that clover creates encourage heavy nitrogen-using plants like wheat, corn, potatoes, and cabbages to grow strongly but will suppress the vigor and growth of legumes. Over time, the pathogens that attack legume roots build up to the point where legumes grow poorly due to root diseases unless species that suppress those specific pathogens, such as brassicas or sorghum, are included at the proper point in the cycle.

Rotation Timing

Timing is everything. After soil building, it is the single most important factor in getting good results on the farm.

— *CR Lawn*[6]

Coordinating planting and harvest, seeding of cover crops, and spreading fertilizers into the whole-farm cropping system is the foundation of the whole-farm operation. The specific timing and related details need to be adapted to local conditions that vary as you move north or south.

Each grain has an optimum planting time that varies with location and can be determined by observing the local grain crops grown in the area and the dates when they are normally seeded. In most temperate climates, both spring and winter grains can be grown. Individual grain crops are better adapted to the soil conditions produced by some species than by others. Also, the timing of seeding and harvest is more compatible between some crops than others. In much of Europe and North America, winter wheat and its related species fit in very well after spring-seeded oats. Growing oats provides an opportune time after harvest for preparing the soil for and seeding wheat. It reduces pressure from diseases and gives plenty of time for applying fertility amendments before winter wheat is planted. Other winter grains, including barley, rye, and triticale, can be grown following wheat, but wheat should not follow other winter grains. Other crops that can precede winter wheat are peas, snap beans, or other vegetables, and short-season soybeans in areas that have a long enough growing season.

Hulled wheat such as emmer, einkorn, zanduri (*T. timopheevii*), and spelt are more tolerant of late planting, poor soils, and low fertility than modern wheat. When hulled wheat is planted relatively early or in rich soil, it produces thick, heavy vegetative growth with lots of straw but has low grain yields. The grain quality is lower from spelt that has been planted early. Late planting produces grain-bearing tillers after growth resumes in the spring and grain with higher protein level and better baking quality.

FALL VERSUS SPRING TILLERING
ON STRESSED PLANTS

Klaas Martens, my trusty einkorn grower of Lakeview Organic Farm in
Penn Yan, New York, wrote the following observations in the summer
of 2015:

Eli,

Your winter einkorn looks good now in early summer. It survived
the harsh winter but had very few tillers, and winter damage in areas
that were most exposed to wind. Less exposed areas came through the
winter with more early growth. Our North Dakota spring einkorn
that was planted last fall looked better early in the spring, but it is
not a good comparison being in different locations with different soils
but it did bring something significant to my attention. We need to
understand more deeply the dynamic of fall tillering vs spring tillers.
Your einkorn looked sparse in the most exposed spots early on but has
filled in remarkably well with huge numbers of spring tillers. In more
protected areas, yours had more fall tillers as it started growing this
spring. It may be that the plant intelligently conserved its resources
when under stress, then in more favorable conditions was able to
flourish as a result.

The sparse-looking stands in early spring later explode with new tillers
and strong growth making higher-yielding grain with better baking
quality than those with extensive fall development. Moderately late
planting makes thin-looking stands going into winter that become
stronger, with higher-yielding crops in the summer. The difference
seems to be in the allocation of resources within the plant, just as later
planted spring grains make taller, bigger stalks with much smaller
heads than early ones. That is related to day length.

My observation is that barley yields well only on fall tillers, whereas
wheat yields are better with the fall tillers but it can do well on spring
tillers. The spring-planted grains may test higher in protein. Spelt
produces impressive amounts of straw with relatively low grain yield

from fall tillers but the opposite on spring tillers. Spelt yields poorly on its fall tillers and has low baking quality while its spring tillers produce both good quality grain and high yields.

Rye can survive late planting but yields by far the best when planted very early and thin in the fall and then sets profuse numbers of tillers followed by additional spring tillering. While it tolerates low-fertility soils, it has the greatest response to high fertility of any of the winter grains. The big caveat with rye is pollination. It is very self-infertile, and only varieties or hybrids with a broad genetic base will make their potential number of grains. It also does poorly when planted as deep as we normally plant wheat or other grains.

Rye is closely related to an older ancestral species that was a perennial and it has two perennial close relatives that also produce grain. Its ancestral "wild" traits must be taken into account when it is grown.

With all of that in mind, I need to observe einkorn more closely. It may be partially perennial or may have vital genes from its perennial ancestors. Its growth habit is very likely to be influenced by day length more strongly than its modern relatives. The North Dakota farmers plant their einkorn in the spring at very high seed density compared with how we do it. They also have high soil fertility where they grow einkorn. All this raises more questions to study than it answers. We need to experiment with more planting dates in the fall and experiment with different seeding rates especially in the spring to observe how ancient einkorn responds to its environment. Einkorn may have very different cultural requirements than I had assumed in the past. Much of what I originally learned about grain production was based on very highly bred modern species while einkorn behaves more like its wild ancestors. Einkorn has to teach me how it wants to grow instead of me making it fit into how I want to grow it.

Have a good day.

Klaas

Increasing Beneficial Mycorrhizae through Rotation, Reduced Tillage, and Landrace Wheat

In healthy soil, you cannot tell where the plant ends and the soil begins.[7]
— *Rudolf Steiner*

Mycorrhizae are good for wheat and good for your soil. These tiny hair-thin fungi have a mutually beneficial relationship with the plant by growing extremely fine, rootlike filaments that reach far out into the soil, acting as root extensions to bring in nutrients to the plant, while receiving in return the sugars that nourish them. A good friendship here. Mycorrhizae increase the nutrient absorption of the plant from a hundred to a thousand times, enhance the uptake of macro- and micro-nutrients, increase wheat's hardiness during water scarcity, decrease lodging in heavy rains, and build soil by breaking down rocks into absorbable minerals. Yield increases while fertilizer inputs decrease. According to researchers who have investigated the role of mycorrhizae on wheat nutrient uptake, "Wheat biomass and grain yields were significantly higher in plants inoculated with mycorrhizae than non-mycorrhizal plots irrespective of soil moisture. This demonstrates the potential for improved growth, yield, and nutrient uptake of wheat with mycorrhizal associations and can reduce the effects of drought on wheat grown in water-stressed field conditions."[8]

How can we manage our whole farm to encourage the growth of these tiny helpful allies? Three interacting aspects include:

1. Keeping your fields in a rotation without fallow or bare soil. Harvesting a crop often brings the death of the plant, or the crop reaches its maturity and ceases to grow. The crop roots die, eliminating the vital food source for the fungi. It's best to under-sow with a cover crop prior to harvest so that there is a continuous living habitat for the fungi.
2. Minimizing tillage and keeping fields in cover crops. Plowing destroys the fungi, but not tilling by itself does not help mycorrhizae thrive. Growers observe that mycorrhizae increase with less tilling. Even in the case of not tilling at all, you can lose the mycorrhizae if there is no cover crop with vital plant roots for symbiosis.

3. Growing landrace wheat. Mycorrhizae like landrace wheats. In an experiment to determine the best crops to encourage mycorrhizal growth, it was discovered that not only is wheat a good habitat for the mycorrhizal fungi, but that landrace wheat is the preferred type.[9]

There is a genetic basis for differences in mycorrhizal dependence among wheat cultivars. Industrial wheat breeders have not considered the interactions between plants and soil microflora as a goal but have focused on dwarfing wheat plants to better adapt to synthetic agrochemicals. Landrace wheats evolved in low-input soils over millennia, developing survival mechanisms for rigorous mycorrhizal associations.

Mycorrhizae love einkorn. Although modern wheat is a heavy feeder, einkorn's uniquely strong mycorrhizal association builds soil. When I pull up an einkorn plant by the roots, it is covered with extensive mycorrhizae. The largest einkorn plants have the greatest clumps of mycorrhizae. Wild emmer wheat, the ancestral progenitor of all wheats, also has strong root-mycorrhizal associations, to the extent that einkorn and emmer are breeding sources of mycorrhizal associations. I have never seen any mycorrhizae on modern wheat roots.[10]

MIXTURES

Looking at the field of ripening grains, Vavilov realized that it was not a uniform wheat cultivar, but a panoply of intermixed strains of grain that formed a resilient polyculture. It was necessary to collect hundreds of seedheads for a representative sample of the vast biodiversity in a single field.[11]

— GARY NABHAN

I spent years collecting landrace wheats in the harshest environments. Yusef, a Bedouin wheat farmer in the parched Negev desert of Israel, explained to me, "Each plant is different. I like the taste of this one to save for seed," as he plucked a handful of seed to munch and offered me a taste. "The plants with big heads have deep roots to reach down for the water. They are good ones." The seedheads of these wheats were the size of small corncobs! I counted forty-eight different landrace strains growing together in Yusef's field.

In contrast to the pedigree wheat monocrops of today's wheat fields, from the earliest days, traditional farmers have grown mixtures of diverse wheat landraces. Fields of wild wheat diversity evolved into landrace metapopulations with dynamic gene flows between diverse wheat plants. Traditional mixtures of landrace varieties mimic the genetic diversity of nature, which contains a vast diversity of genotypes. Landrace wheats are typically composed of mixtures where one genotype dominates and others, including natural hybrids, occur. Traditional farmers typically combine diverse varieties together in proportions differing from farm to farm and village to village. Combining mixtures of different varieties with the same maturity period creates a natural buffer against diseases and pests, enhancing the populations' evolutionary capacity for resilience and complexity of flavor.

A study on mixtures documented their potential to outyield a monoculture.[12] This evolutionary process ensures yield stability over years. However, lack of breeding knowledge or availability of quality varieties can limit substantial improvement of landraces on traditional farms in remote regions. Stable yields in organic conditions favor the polygenetic traits of landraces over modern pedigree uniformity. Genetically diverse populations allow for adaptation through self-regulating, evolutionary natural systems that generate flexible, adaptable traits. The practice of growing mixtures not only decreases the risk of crop failure and enhances food security but imparts a richer, complex flavor to breads and beverages.

In 1880, Philippe-Victoire de Vilmorin, the master seedsman of France, reported:

The planting of mixtures of two or more varieties to achieve a more abundant and assured harvest is recommended by experienced farmers, and we believe commendable indeed. It is well established by experiments that mixtures of two distinct varieties of wheat yield almost constantly greater than one variety alone. Thus we often see farmers plant their crops skillfully with wheat mixtures. Each variety of wheat differs subtly from all others, not only in its external characteristics, but by its manner of drawing nourishment from the soil. These slight differences have a significant impact on overall performance. If one plants the seeds too tight, the wheat is a formidable weed against itself. This is even more so if all the feet that are struggling and competing belong to the same variety, because the roots of each find themselves in constant contact with the roots of similar plants at the same time and at the same depth, seeking exactly the same foods. However, if different varieties are planted together, the competition is not as intense. Almost always one variety will dominate, so it is good to grow each separately and purely, then mix them together at the time of sowing, in the proportions to be most advantageous. Care must be taken to grow wheats that mature at the same time.[13]

According to the Georgian wheat expert Dr. Taiul Berishvili, "Since the dawn of agriculture, mixtures have been sown to enhance the complementary growth habits of various types of wheat in the Caucasus center of wheat domestication. This is not a simple mechanical mixture but rather represents a complementary relationship between diverse wheat species or rye and barley in the taking up of nutrient substances from soil, as well as access to water and light, the result of which is their successful cohabitation. This practice suppresses disease and increases yield. Traditional mixtures in Georgia include:

Persian[14] *and soft wheat*. This mixture was planted to give the grains better baking quality.

Persian and barley. This mixture was combined for its good performance under high-mountain, cold-climate conditions, where Persian followed barley, which ripens earlier. Barley is short, and Persian wheat is tall. Their complementary heights permit a dense spacing that lets the light into both plants and increases yield. Bread baked from a Persian-barley mix is preferred in quality to barley bread alone.

Wheat and rye. Rye naturally infested wheat crops as a weed in earlier times. Farmers observed that in harsh weather the rye would survive even when the wheat crop failed. This encouraged Georgian farmers to plant wheat-rye mixtures intentionally. In this case, rye occupies the highest leaf canopy because it is taller than wheat. Bread baked from this mixture was known as maslin in Europe.

Wheat, rye, Persian, and soft wheat. Both the soft wheat and dika were found in the wheat-rye mixture. All this is evidence of the traditional Georgian farmer's keenness of observation over generations.[15]

SECRET OF THE DEPTHS:
WIDE SPACING IN LIVING SOIL

*We plant landraces farther apart to accommodate larger root systems and
greater height that impart rich flavor and decrease disease.*

— GLENN ROBERTS

Seeding rate, seed spacing, seeding dates, soil fertility, and tiller numbers
all interact together to determine the number of heads per square yard.
Final stand density plus head size and kernel size are key factors that
determine yield.

Seeding at the appropriate time encourages robust early root growth
for vigorous plants. Seeding depth and spacing are more important than
most growers realize. Planting small grains too deep diverts energy from
early tillering and extends the length of the hypocotyl. When this happens,
the stand may look fine, but density and yield potential have already been
severely limited. Maximum seeding depth for oats, wheat, and barley is
about 1¼ inches; for rye, einkorn, and emmer, not more than ¾ inch deep.
Plant slightly deeper in sandy soil and shallower in clay soil. Seeding rate
should be calculated starting with the optimum final stand. If we are trying
to establish a stand of fifty-five heads per square foot (a good target for
many varieties of wheat) consider planting date, soil fertility, species, and
fertilizer rate to estimate the likely degree of tillering.

The number of tillers the wheat plant puts out is determined earlier than
most farmers think. As young seedlings interact with the soil, they generate more tillers if there is ample root space, good tilth, and soil nutrients.
Tillering takes place even before flag leaf is visible. The root system belowground reflects the tillering aboveground. By understanding this basic habit
of the wheat plant, a farmer may harness tillering capacity to great advantage. Poor fertility or crowding in infancy can never be compensated for
later. Stress such as drought or nitrogen deficiency at this vulnerable stage
results in irreversible loss of grain yield. Modern wheat generally has one
tiller, whereas landrace wheat may have as many as forty to fifty tillers, each
with a seedhead when provided with ample nutrients, root space, and depth
around each seed. Einkorn may grow twenty-five tillers or more per plant.

Why is wide spacing so important? Wide spacing promotes more extensive root systems; strong, active stalks; and greater disease resistance. The

nutritional balance that gives all plants more active immune systems is rarely if ever considered in the conventional fertilizer industry. The landrace grains that naturally grow with long straw, in particular the winter grains, are more capable of metabolizing a wider range of mineral nutrition to build the cell and seed structure. Their stronger, more adaptable root system has a significantly greater surface area under the ground to host more beneficial soil biota as partners.

Wide spacing is an important strategy to mitigate the stresses of climate change on wheat crops. It stimulates deep, extensive root growth that enables the plant to better survive drought, heat, and rain extremes. After three years of selecting wheat in wide spacing, we have observed an increased adaptability to weather extremes as measured by less lodging under heavy rain and greater tolerance to drought stress as measured by leaf color and overall plant robustness, and stable yield with protein levels comparable to commercially grown grain varieties.

Most wheat is planted at a higher than optimum seeding density. Thirty to forty-five heads per square foot is a high stand density that may be appropriate for intensively managed modern wheat in a high yield environment, but the same or greater number of large seedheads can grow from one robust landrace wheat planted at 12-inch spacing.

Ancient and heritage grains tiller far more prolifically than modern wheat. Twenty-five to forty tillers per einkorn plant is not unusual. Fifteen to twenty-five tillers per emmer plant is common. Most heritage wheat varieties can easily set five to fifteen tillers per plant with good growing conditions, and up to forty or even fifty in optimal conditions. This illustrates why seeding rate should be reduced when growing older types of grain.

The key factors that interact to determine the number of productive tillers per plant are timing, spacing, and soil fertility. An earlier seeding date will increase the number of tillers. Greater distance between plants increases the number of tillers. Higher fertility increases the number of tillers. Faster and stronger emergence also increases tillering.

Typical seeding equipment may leave areas with below optimum stand density, or even bare spots, while other rows are too crowded. When seeds grow in clumps, the size and vigor of all the affected plants is substantially reduced. It is very hard to manage a stand of grain with widely varying populations. Farmers generally use higher seeding rates, especially with modern varieties, because the perceived yield penalty for not having enough stand density is greater than that for having too high a population. It's easy to see that thin stands and low populations result in lower yields. Low stand density can be caused by uneven seed spacing, uneven seeding depth, variations in firmness of the seedbed, disease pressure, or fertility imbalances. Excessive density causes a greater risk of lodging and increased disease pressure. Wheel tracks are often so tightly compacted that seed remains on top of the ground, and several rows beside each other have serious gaps. Seed depth often varies greatly, especially with nonuniform soil preparation, adding to the variation in growth and vigor. Uneven seed spacing results in heavier weed pressure that reduces yields.

Carefully observe the size of your seed if you have been selecting for dominant seedheads in the seed-saving stage. The larger seedheads produce larger seeds that may not drop as easily in a mechanical planter, which will then inadvertently select for smaller seeds that drop through more easily.

Spacing in 1868

In the 1868 *The American Wheat Culturist,* S. Edwards Todd explains,

There is no purpose in sowing more than is necessary for the desired goal. That is all that a farmer can expect his seed to do. The struggle for existence among excessively thick sown plants will inevitably lead to the partial starving of all and the deterioration of the seed quality, so that even if the harvest be large in the first season, it will cause subsequent diminished yield and disease arising from weakened seed in future seasons. Each variety has its own optimal spacing within this range, which may be learned through spacing trials.

In 1868, Mr. J.P. Nelson sowed 11 lbs of wheat evenly on one acre. He reports, "The wheat grew luxuriantly beyond anything I ever saw, at least 40 stems each with good heads from one root. The wheat was reaped and thrashed in June, 1869. It yielded seven and a half bushels of the best seed I've ever seen." Although the seeding was excessively light compared to typical rates of today, the yield was quite above average. A lighter seeding rate not only gave the largest yield, but the finest quality. It was by far the heaviest in weight and had the least disease.[16]

The common seeding density crowds and damages the wheat plant. Typically a farmer will plant one bushel of wheat (between 54 and 60 pounds) per acre. This will give 16 grains per square foot, or one kernel per 3 square inches. This is too close for winter wheat, which, due to its longer period of growth, sends out from thirty to forty stalks or even more with ample spacing in good soil. The seeding rate of one-quarter of a bushel (12.5 pounds) to one-third of a bushel (18 pounds) of good seed on an acre is the maximum density for healthy, high-yielding landrace wheat plants.

The SRI (System of Rice Intensification) International Network and Resources Center researchers have conducted extensive international research that confirms that wide spacing significantly increases yield and crop health.[17]

Northeast Organic Wheat Spacing Trials

In randomized, replicated trials with six landrace wheats, we planted at 12-inch, 8-inch, 7-inch and 6-inch spacing, as shown in table 3.2.

Our seed-saving method involves planting 12 inches between each seed (5 pounds per acre), selecting the largest elite seedheads from the most disease-resistant, robust seedheads, repeating this process for two or three years till ample seed is generated, then planting at 8-inch spacing for field production. Planting at a spacing denser than 8 inches between each seed (more than 12 pounds per acre) decreases yield per landrace plant. In my spacing trials, at 6 inches between each seed the plants were an average of 5

Table 3.2. Seeding Rates for Heritage Wheat

Purpose	Spacing
Seed saving: largest seedheads but the plant produces new tillers till it dies, giving 10 percent immature seedheads. Winter wheat.	12 inches between each seed (5 pounds per acre). Optimal for seed saving of the largest seedheads, harvested by sickle or scissor.
Field production: for flour, highest yield with even heads. Winter wheat.	8 inches between each seed (12 pounds per acre).
Spring wheat: good yield, fewer tillers, 3 inches shorter.	7 inches between each seed (16 pounds per acre).
Crowded: Plants 5 inches shorter, fewer tillers, smaller seedheads, more disease, lowest yield.	6 inches between each seed (20 pounds per acre).

inches shorter and had an average of fifteen fewer tillers, each with smaller seedheads. Protein at different spacings was not measured.

Nineteenth-Century Wheat Seed Spacing Trials

Our experience is confirmed by an experiment on wheat seeding rates published by the USDA in 1869 with Tappahannock wheat:

1. 14 pounds per acre planted / 3,456 pounds harvested / highest quality ever seen

Drilled September 22, 1868, on rich, well-drained clay soil, at the rate of one peck (14 pounds) to the acre. This rate yielded 52 bushels per acre. The grain was superior to any other wheat heretofore grown in the county. It weighed 64 pounds per measured bushel.

2. 54 pounds per acre planted / 2,299 pounds harvested / superior quality

Broadcast on the same date on similar soil at the rate of 1 bushel per acre. This yielded 38 bushels per acre, weighing 60½ pounds per bushel. It was superior to ordinary varieties.

3. 108 pounds per acre planted / 873.75 pounds harvested / good quality

Broadcast on the same date on similar soil at the rate of 2 bushels per acre. It yielded at the rate of 15 bushels per acre, each weighing 58¼ pounds per bushel. The grain was the same quality as the best summer varieties.

It will be noticed in this experiment that the lightest seeding rate not only produced the largest yield of grain, but also the finest quality and by far the heaviest in weight.[18]

In *The American Wheat Culturist*, S. Edwards Todd shares correspondence with J. J. Mechi,

J. Mechi of England, who has had much experience in growing wheat, writes:

"one kernel in a hole, at intervals of nine inches by four, would, under favorable circumstances, be ample and produce much more than if four times that number were sown. . . .

Thin sowing is the first cause of large and vigorous ears to select from. On this point, there can be no mistake, seeing that thick sowing has an exactly reverse effect, diminishing and crippling the growth of the ear until, with extreme quantities, there is scarcely a good kernel or a good ear. Therefore, in order to get good ears to select from, or to be certain of the largest possible yield of grain, sow only a moderate quantity per acre. I think that every intelligent wheat grower will agree with me that thin sowing has quite as much or more to do with a large product of superior grain, as the choice of a prolific variety. No rule can be laid down that will serve as a reliable guide for farmers in various portions of the country in determining the quantity of wheat per acre. For this reason, I shall not attempt to state how much this farmer, or that wheat grower, should sow per acre.

In a letter dated June 27, 1867, Mr. Mechi says:

I related last year that a peck [14 pounds] of seed wheat per acre, dibbled at intervals of about 4½ inches, one kernel in a hole, produced 58 bushels of heavy wheat per acre . . . in fact, the thickest and heaviest crop of corn [wheat] and straw on my farm. It was seen at various periods of its growth by many agricultural and other visitors. During winter, a single stem only having appeared from each kernel, the land at a distance appeared as if unsown, and we were often asked why we had omitted to drill that particular portion of the field. In the spring each stem radiated its shoots horizontally, to the extent in some instances of thirty to forty-eight stems, and ultimately became the best crop on the farm, and, which is often convenient in harvesting, about four days later than the thick sown put in, in October, at the same time as the rest of the field was drilled with one bushel per acre. In October last, rather late in the month, we repeated the experiment on a heavy-land clover lea, as last year. The ground was rough and hard, and very dry, and although a kernel was placed in each hole, only about one-half, or half a peck per acre, came up. Of course we anticipated a partial failure, but spring came, and each stem threw out horizontally a large number of shoots, so that now it is admitted by all who see it that it will exceed in produce the adjoining crop, drilled at one bushel [56 pounds] per acre.[19]

The recommended seeding rate in 1869 is consistent with the results of our research today. Eleven pounds per acre is a spacing slightly wider than 8 inches between each seed. Specific seeding rate recommendations vary from variety to variety, so therefore we advise that each farmer conducts on-farm trials to determine optimal seed rate for a variety according to your soil and fertility management practices, and to select for preferred traits.

1880 Seed Density Trials

In France in the late 1800s, the Vilmorin seed company reports on their seeding rate trials:

Among many experiments we have made on this subject, we will mention one that is conclusive. In a field of good soil under ordinary conditions of wheat culture, we planted a winter wheat in the month of October in four plots of equal extent. One of them that served as comparison had 180 liters of seed per hectare [one hectare equals 2.47 acres], while others received only half, the third, and the sixth seed to the first, that is to say respectively 90, 60, and 30 liters. We found at harvest the yield of straw and grain increased as the rate was more lightly seeded. Not only was the performance of the last at 30 liters the greatest, but the grain was the best and heaviest at the same volume. Later in season it can be sown thicker. Size or fineness of grain must also be considered. Experience is the best as a guide, but it must be informed by thinking and reasoning.[20]

Vilmorin goes on to explain that the seeding rate of wheat sown by one farmer cannot be a practical guideline for another unless the variety, soil, and time of seeding are similar. Since kernels of wheat vary in weight and size, the number of grains in a pound will vary.

Cutting Back Wheat in Spring
to Prevent Lodging

The Vilmorin book reports, "Here is a useful remedy to prevent lodging. If a wheat grass becomes too vigorous in fertile soil, to fortify the plant against the danger of lodging, farmers cut back the leaves with a sickle or scythe. The stems will be strengthened against lodging by the stimulation thereby. The danger of lodging will be lessened, if not altogether eliminated."[21]

Seeding Depth

Seeds should be planted at an average depth of 1 to 1¼ inches, slightly deeper in sandy soil. Shallower depth may cause the germinating seed to dry out. Deeper than 1 inch decreases germination and emergence rate in heavy clay soil. In addition to the lower population density and more even seed spacing we recommend here, seeding depth is an important factor in stimulating strong plants for heavy tillering. Maximum depth for small grains such as oats and wheat should be 1¼ inches. With rye, this needs to be even shallower: ¾ inch is the maximum seeding depth. Seeds planted deeper than ¾ inch will divert energy and growth from early tillering into producing a longer hypocotyl for timely emergence from the soil, causing reduced yield potential due to fewer tillers. Seed depth is especially important at lower seeding densities where heavy tillering is needed to produce thick stands.

Irrigation

Glenn Roberts of Anson Mills explains:

We never irrigate landrace crops, with the exception of rice, of course, which is why we have about fifty percent success in any given year. Irrigation suppresses flavor in landrace crops. Climate stress improves flavor in landrace crops. All of the above practices we follow in landrace crop farming are exactly the same as, or simulate to a significant degree, horticultural methods from antiquity.[22]

COMPOSTING FOR LIVING SOIL FERTILITY

Well-mineralized soil, rich with organic matter, has a natural resilience to pests and disease.[23] How can the inert elements in the soil and atmosphere be biologically enlivened to transform into the tissue of plants? Plants are better able to absorb soil amendments that are biologically processed in an active compost pile, tilled under with cover crops or mixed with almost finished compost to allow soil organisms to enliven them. Heritage grains respond well to the addition of enlivened mineral amendments since they have a greater biological capacity to host more mineral transferring soil organisms that reach a wider range of elements than modern grains with stubby roots.

Cool Compost Optimizes Beneficial Soil Organisms

Our goal is to create a compost teeming with life. Active, complex communities of beneficial soil organisms are a powerful inoculant to nourish soil life. Cool-managed decomposition mimics the processes of nature, encouraging a greater biodiversity of beneficial compost organisms, particularly fungi and earthworms. Excessively turned thermophilic ("hot") composts have fewer beneficial organisms. Turning inhibits the full development of the needed fungi in their mature sporulating stage. Living soil interactions are our original wealth: in Hebrew, Adam (Adama) means soil and Eve (Havah) means life.

Soil Structure and Tilth

A well-aerated soil provides an optimal environment for enlivened amendments that can be available to the extensive root/mycorrhizal range of landrace wheat. Heavy machinery tills soil without consideration of compaction, limiting the plants' capacity to utilize elements accessed by soil organisms. The soil below the aerobic zone is generally not serving the crop when compacted by machinery.

Basic Guidelines for Amendment-Enlivened Compost

A mesophilic, cool-managed grass-clipping, leaf, and manure compost should be well mixed from the start with healthy soil, humates, and Azomite to increase the potential soil fertility that is; the cation exchange capacity.[24] Mix in powdered clay if your soil is sandy. Incorporate horsetail (*Equisetum arvense*) for the homeopathic-like effect from vegetative silica. The pile does not need to be turned. Cover the pile and locate it in a shady section of your field to maintain the moisture that increases microbial diversity. If it dries, it dies.

Farmers can make their own unique compost amendment recipe to address the strengths and deficiencies in their fields. Incorporate rock powders toward the end stage when the compost is not hot, so that it does not depress the decomposition process, which prefers a slightly acid pH. Approximately 50 pounds of mixed rock minerals added to a cubic yard of half-finished compost is good. The compost may be mixed with the seed at planting time, applied to the rows of sown grain, or cultivated into the soil.

Test Your Soil, Test the Test

Even if soil tests indicate that the necessary elements are in the soil, they may not be biologically available to the plant. A revealing experiment to discover the diversity of soil-testing methods is to take a well-mixed soil sample from a single field and send a portion of that identical sample to different soil labs for analysis and see the disparity between the results, as well as the recommendations that may accompany them.

Plant Indicators

A plant community acts as an indicator of the environmental conditions of its habitat, telling the trained observer the soil fertility, nutrient deficiencies, water saturation, compaction problems, pH levels, and more. Perennial weeds that colonize an area over a long period of time may be a more reliable form of indicator than annual weeds that may indicate a temporary condition. Mark Fulford observes, "Living indicators give compelling information on how to grow superior grain crops. Northeastern soils tend to grow broadleaf weeds and annual grasses in tilled ground." In *Weeds and What They Tell*, the biodynamic researcher Ehrenfried Pfeiffer says, "Fertile soil supports plants such as Stinging Nettle, Lambs Quarters, Pigweed or Purslane."[25] Plantain may indicate soil compaction. Sorrel or brackens like acidic soils. Alkaline soils tend to support populations of perennial Sow Thistle, Bladder Campion and Henbane. Perennial indicator plants in rundown pastures and hay fields include Bed Straw, Goldenrod and Queen Anne's Lace. Farmers need to observe for themselves the environmental indicators in their own field. If a soil test agrees with the evidence expressed on the land and the indicator plants, then it is likely to be a useful test.

Leaf and Sap Tests

Use leaf and sap tests[26] to assess the elements that are in the leaf and petiole. Read the color and structure of indicator plant populations and assess their relative robustness. Flavor is an important indicator as well. If stems are flimsy or roots discolored — not firm and white — this is a message from nature to change. Practice fertility methods to increase aeration, soil biology, microbial foods, and the wide spacing of plants.

MINERALS AND MICROBES

Fertility deficiencies and imbalances are related to the microbial activity in the soil. Microbial activity increases as the soil warms and is much greater in moist aerobic conditions than in waterlogged anaerobic soil. Phosphorus availability is especially affected by microbial activity. Phosphorus is important in the transport of energy within plant cells. Adequate levels of phosphorus are important in early seedling development, in tillering, and for good winter hardiness.

In a two-year test funded by a SARE grant, in which soil samples from seven different fields in different stages of crop rotation were taken every six weeks, Klaas Martens found that the availability of phosphorus varied with the time of year and weather conditions. Phosphorus levels found in soil tests were lowest in February and generally peaked in July or August, then declined rapidly through the fall. This is extremely important when growing heavy-feeding crops with relatively weak root systems like modern wheat. The demand for phosphorus in grains is highest during the six weeks after emergence; that is, just as phosphorus levels are declining in the fall and before they have begun going up again in the spring. Cold, wet soils can delay the mineralization of nitrogen as grains come out of dormancy in the spring. In soils with low phosphorus-supplying ability, it may be necessary to apply materials high in phosphorus, such as poultry litter, in the fall ahead of planting winter grains.

Our atmosphere is 78 percent nitrogen, but without the amazing work of invisible soil organisms, gaseous nitrogen could not be transformed into the proteins that sustain plants, animals, and people. Nitrogen depends on the activity of nitrifying and denitrifying bacteria to be available to plant crops. Many farmers have observed that nitrogen from nitrate (NO_3) tends to stimulate vegetative growth in grains while nitrogen from ammonia (NH_4) promotes maturation, because nitrate stimulates cells to divide and plants to grow rapidly. When temperatures are low and soil is excessively wet, microbes mineralize nitrogen into nitrate slowly, while excess water tends to leach it away. When soil is waterlogged, anaerobic conditions develop, which favor denitrifying organisms and discourage the aerobic nitrifying bacteria. In both excessively wet and dry conditions, plants have difficulty taking in nitrogen from ammonia and urea sources; thus, having a sufficient supply of nitrate nitrogen in the soil is more important during times of extreme dry or wet conditions.

When nitrogen mineralizes and becomes available to the crop is as important as how much is available. Fortunately, well-balanced organic systems usually mineralize adequate quantities of nitrogen at a rate that matches the needs of grain crops. If too much nitrogen becomes mineralized at one time, the crop is affected as much as if too much synthetic nitrogen fertilizer were applied at one time. The stalks grow too fast and are prone to lodging and disease. If it cannot mineralize rapidly enough during peak demand periods, the yield and protein content may be reduced. If excess nitrogen becomes available before or after the crop can utilize it, weeds will proliferate.

Good soil structure and adequate drainage help maintain soil conditions that are favorable to aerobic organisms during periods of excessive precipitation. They also keep the soil favorably moist during droughts. When soil has high levels of organic matter and plenty of high-quality organic residues, organisms like earthworms, ground beetles, and protozoa thrive, and the rest of the soil food web is healthy. Active soil biology is essential for mineralizing nutrients and cycling them rapidly enough to meet the needs of high-yielding crops.

SILICA, SPACING, AND LODGING RESISTANCE

Plant-to-soil organisms communicate by way of hormones, chemical exudates, and subtle electromagnetic impulses inherent in all living beings. Trigger events such as stress from insects, disease, or weather signal the mutually supporting soil organisms (a consortium of mostly fungi and associated bacterial colonies) to alert plant defense mechanisms. In mere hours, a crop can alter itself physiologically to develop more fiber and disease-resistant components. For example, mowing or grazing winter grains in early spring stimulates greater tillering and decreases lodging potential. Enhancing silica in the soil promotes subtle shifts that may be the difference between a successful harvest and crop failure.

How can farmers increase the silica in grains? Harness the soil biota. Feed the soil life. Set the stage for the highest possible soil-to-plant relationship. Silica is in all soils, especially in the colloid particles of clay soils. Silica promotes a greater cation exchange capacity by increasing the amount of root surface area to accommodate more ions of elements for exchange. The ability of crops, especially those in the grass family, to gain access to silica is primarily through the mycorrhizae and, to a lesser extent, via bacterial colonies associated with the roots. Inoculation can help when soil biology is sluggish. If you purchase a commercially available mycorrhizal inoculant, make sure there are actinomycetes, streptomycetes, or both listed on the label, not just mycorrhizae. It helps to have phosphate-reducing bacteria as well, since phosphate easily locks up iron and calcium in soils of low biology and humic forms of organic matter. Silica, as well as calcium, needs the help of humus, phosphorus, boron, and fungi to become available. Aeration is critical, as soil gas exchange is often the reason silica-trading soil microbes do not develop well. They are oxygen-dependent just like humans.

Widely spaced grains grown with ample silica have sharper edged leaves compared to the limp, soft, conventionally grown grains. Stalks become remarkably stout yet flexible, not prone to collapse in heavy rain and wind damage. This is silica in plants at its best. Landrace grains are capable of developing this quality much more so than short-bred (pun intended) strains that have disproportionally heavy heads compared to the weaker backbone of the straw needed to support them.

See for yourself: Cut a cross section of a well-grown landrace wheat straw above the last flag leaf, right up to the grain head. You will see what is rarely seen these days: a solid, stemmed straw, filled with a stiff white cellulose (almost like styrofoam) that prevents the stem from folding and kinking as it nears ripening. It is primarily composed of silica, pectin, and cellulose.

Incorporating finely ground or water-soluble humates into your compost or tilling in a cover crop spread with dry minerals is an effective way to "carbonize" (humify) minerals, rock dusts, and volatile products, especially those that were alive in their last lifetime such as ground shells. This can prevent reactive minerals that don't store well together such as calcium and phosphorus from becoming tied up before their useful time of application. The gypsum will not raise pH and shut down the heating cycle. It also lends much-needed sulfate for the rapid reproduction of bacteria. Mixing these relatively common rock powders with good-quality compost can put a new level of energy into farm soils.

Biodynamic Foliar Applications for Plant Support and Disease Control

Biodynamic (BD) horn silica spray[27] introduces silica and other beneficial elements that not only strengthen stalks but help prevent moisture-related diseases such as fusarium or mildews. Foliar spraying is best done in the most receptive times: still, early mornings; evening hours as the sun is going down; and even at night, when plant stomate cells are open. The soil is exhaling more carbon dioxide than in the day, the cell walls are more receptive to penetration, and the above-listed liquid spray can also be combined with other adjuvants such as diluted raw milk or diluted seaweed extracts. Not all of these are compatible with each other, so one must do jar tests to make sure all components stay in solution. A backpack sprayer can be used to apply liquid silica as in Sil-MATRIX from PQ products, BD horn silica spray or horsetail teas, with solubilized fulvic acid in the mix.[28] It is a small and slippery chelating molecule that brings ions of other elements directly into the leaf.

Nature Farming and Double-Harvest Low-Till Wheat

The uncultivated land surrounding the farmers' fields do not show signs of trouble. The plants growing in the fence rows thrive through drought as well as in fine weather. Would that observation justify us in wondering whether the manner in which farmers handle their land might be responsible for the stressed crops grown under tillage? May we suggest to investigate whether men could grow healthy crops if they copied the conditions which prevail in nature where crops are universally healthy?[29]

— EDWARD FAULKNER

Masanobu Fukuoka, author of *The One Straw Revolution*,[30] developed a system of natural farming without tilling the soil, but with generous natural mulching and the extensive culling out of weaker plants. He did not plow his fields, used no agrochemicals or prepared fertilizers, did not flood his rice fields as farmers have done in Asia for centuries — yet his yields equaled or surpassed the most productive farms in Japan. How did he achieve this? In the words of Larry Korn writing about Fukuoka:

Look at this grain! A revolution can begin from this one straw. It is simple: To plant, I broadcast rye and barley seed on separate fields in the fall while the rice is still standing. A few weeks later, I harvest the rice and spread the straw back over the fields from which it came. It is the same for the rice seeding. About two weeks before winter grain (rye and barley) are harvested, I broadcast rice seed over them. After they have been harvested and the grains threshed, I spread the rye and barley straw over the field. Using the same method to plant rice and winter grain is unique to this kind of farming. Take out the weaker plants and develop strong plants year by year. You may notice that white clover and weeds are also growing in these fields. Clover seed was sown among the rice plants in early October. I do not worry about sowing the weeds. They reseed themselves quite easily. Ha!

The order of Planting: In early Fall, I broadcast clover among the rice followed by winter grain. In early November, the rice sown the previous year is harvested. Then I sow next year's rice seed and lay straw across the field. The rye and barley you see in front of you were grown this way. In caring for a quarter-acre field, one or two people can do all the work of growing rice and winter grain in a matter of days.

With this kind of farming which uses no machines, no prepared fertilizer, and no chemicals — it is possible to attain a harvest equal to or greater than that of the average Japanese farm. The proof is before your eyes.

This is a balanced farm ecosystem. Insect and plant communities maintain a stable relationship here. The reason that modern agriculture seems to be necessary is that the natural balance has been so badly upset beforehand by those same techniques that the land appears to be dependent on them. As Sensei wrote, "Natural farming is not simply a way of growing crops. It is the cultivation of human beings."[31]

The heart of nature farming is to observe your land and crop cycles. *Heavy culling is the key.* Otherwise you will end up with a field of weak, crowded wheat plants.[32] If you are starting the process, in April plant a bed of clover. In the end of August press the wheat seed deep into the soil through the carpet of clover, at a spacing of 12 inches between each seed. During its longer vegetative period the wheat plant will generate deep roots before winter, nourished by the generous availability of nutrients and bacterial activity in the living soil.

Clover is a perennial. It is not sown each year, but at the beginning of the season it is mowed down to a height that does not damage the wheat plants. After harvesting in August, the straw and chaff are spread on the field. In this method, tillage is reduced to a bare minimum, avoiding the plowing and compaction that depress the life process of revitalizing the soil.

There are several advantages to Fukuoka's No-Till Method: in contrast to the conventional way, tillering begins in early September, allowing a longer period for root development; a more active root system in a bed of clover reduces leaching of nutrients in winter; larger plants are more winter-hardy than young seedlings; and, with a wider spacing, more mature tillers are produced, each producing a fat seedhead. One problem, however, is that this system does not work well with modern wheats, which are bred with shorter roots. Long-straw, traditional facultative wheat landraces with strong vegetative vigor and allelopathic root exudates are optimal for this no-till system. Intensive rogueing and culling out of weaker wheat plants is essential.

Mechanical harvesting with a tractor will be difficult with this method since the clover may tangle up the blades. A cradle scythe or handheld sickle works well, since crop residues are cut and left to slowly decompose on the surface of the soil.

In order to control the greater potential for fusarium when wheat residues are left on the soil surface, cover with a layer of shredded leaves or mowed grass as a natural mulch to encourage decomposition. This will suppress weeds without inviting in the fusarium spores of old wheat stalks. Earthworms love this decomposing surface environment. The disease-suppressive earthworm castings will control the spread of fusarium. Alternatively, compost the old stalks, then apply the compost after a year.

WEED CONTROL

Seedbed preparation and seeding the crop is the first opportunity for weed control. Anything that encourages rapid early growth and canopy closure reduces weed growth. A firm seedbed with enough soil moisture to get quick, uniform emergence gives the crop a big advantage. Ideally, the soil immediately around the seed should be firm and moist, while the soil between and over the drill rows should be loose and dry. A big challenge to achieving this kind of soil condition is that many grain drills are pulled directly behind tractors that make deep compacted wheel tracks that several of the seed openers must plant into. The soil in these tracks is very tight and hard, often preventing the seed openers from operating at the desired depth and creating soil conditions that favor weed emergence. The depressions left by heavy tractor wheels are often deep enough to affect the performance of tine weeders and rotary hoes. As tractors have become larger and heavier, this problem has increased.

Despite all of the best efforts a farmer puts in, weeds will grow and must be controlled using multiple strategies. Including small grains in your cropping system is an excellent way to reduce weeds. The rotation will change the soil environment season by season so that no one weed species can become too well established.

Methods for weed control include: plowing under weeds prior to planting, intercropping grains with heavily seeded clover to suppress weed growth, selecting varieties for height to shade out weeds, harnessing allelopathic root exudates combined with closer plant spacing, and crop rotation. Farmers with serious weed problems, such as Canada thistle or field bindweed, should control these weeds prior to planting grain. Alternate field use such as pasture or hay production may be preferable until serious weeds are brought under control.

There are three ways to kill a weed: desiccate it, suffocate it, or cut it so the remaining parts are not viable. Weeding tools can do all three, but the farmer needs to assess the conditions to determine the best strategy for each situation.

In dry conditions, desiccation is generally the most effective way to kill weeds. Burying them is often ineffective when it's dry because they can push back out unless they are covered quite deeply. When it is wet or rain is imminent, burying weeds is very successful as they will become sealed in when being covered after rain falls. Desiccation is not an effective approach in wet weather.

Tools for Weed Management

To get good results from weeding, it's important to select the right tool and adjust it correctly for each unique condition. Achieving excellent conditions to encourage fast and uniform crop emergence can be difficult with a lot of commonly available seed drills. It takes great attention to detail when preparing the soil so that grain can be given the best possible start. Some manufacturers offer harrow attachments for the front of grain drills that loosen the wheel tracks and make a uniform seedbed right ahead of the drill. Some farmers hitch their drills directly to the back of a harrow to achieve a similar result. Most of these approaches improve the quality of the seeding job but can be cumbersome to operate and troublesome to transport. Manufacturers have recently begun offering machines that combine tillage and seeding functions in one pass. These tend to be heavy and require large tractors to pull them, but they do a much better job of seeding than older machines. Ironically, the old grain seeders that were pulled by horses or by very small tractors often were better. To be sure, the most advanced modern machines do an excellent job. Unfortunately, they are expensive and require big, high-performance tractors.

Weeders, including tine harrows, rotary hoes, and even chainlink drags, can do a good job of eliminating weed seedlings from a grain field. Small grains can tolerate an aggressive weeding until the day before they emerge. Most weeds have very small seeds that must germinate in the top half inch of soil. Most crop seeds are larger and are planted deeper. When weeds are in the white thread stage, just prior to or immediately after emergence, and the crop has not yet emerged, there is an opportunity to do shallow, "blind cultivation" with a weeder to destroy much of the first flush of weeds with minimal damage.

With tine weeders, each different type of tine has its own action. Sharply bent tines such as the Lely or Williams system work exceptionally well for weeds that have branching roots. The tines hook the branch roots and pull them up while leaving deep-rooted crop plants, especially those with tap roots, undisturbed. Unfortunately, grains quickly form strongly branching roots. Weeders with sharply bent tines can do serious damage to grain crops and should be avoided except in rare circumstances. Straight tines generally move soil sideways and bury weeds. They work best in flat seedbeds where all of the tines work uniformly. Tines that are bent about 45 degrees at the end follow the contour of the soil well and tend to uproot weeds. The tine angle can be adjusted to make them more or less aggressive, and weight can be transferred to or from the tines. Increasing ground speed makes the action of the tines much more aggressive.

Rotary hoes work best when the soil is slightly crusted and the crust is slightly damp. Under these conditions they break the surface up into small pieces of soil and lift them, an action effective on small weeds yet gentle on crops. Rotary hoes can be adjusted by adding weights to the tool bar and by changing its height. They normally need to be operated at high speed, but changing the speed also changes the action of the tool.

After crop emergence, there is a second opportunity for weeding. After the two-leaf stage of the grain, when the plants have become well rooted, they can be tine-weeded or rotary-hoed once again. Slight damage from the weeders at that point sometimes stimulates stronger crop growth. Many years ago, farmers ran weeders pulled by horses over weak or thin stands of winter grains early in the spring to stimulate additional tillering.

Annual Weeds

Annual weeds seldom trouble fall-seeded grains but can be a serious problem for spring grains. To control weeds in spring grains, establish a strong, uniform stand that can suppress any weeds that get started. Landrace grains have allelopathic root exudates that reduce the growth rate of small weeds. They are also frost-tolerant and can begin growing at lower soil temperature than many weeds. Unfortunately, when grains get off to a slow start, or are planted in widely spaced rows that allow sunlight to reach the ground for a long time, or when they fail to tiller heavily and allow the sun to penetrate the canopy, weeds can get established and cause harvest problems and quality issues. Undersowing with clover is an effective method to outcompete weeds, especially with widely spaced spring grains. Spreading leaf mulch after emergence can help suppress weeds. Make sure to shake off any mulch that may cover the little wheat leaves.

Perennial Weeds

Perennial weeds grow from underground root systems. They are difficult to eradicate when they become well established. They generally have large reserves in extensive root systems that regrow rapidly even when cut above the soil. To remove stubborn perennial weeds, change the favorable conditions that they like. Altering the soil environment is the priority strategy to control them.

Ubiquitous quackgrass, also known as witchgrass (*Elymus repens*), spreads rapidly by creeping rhizomes, which cause heavy competition for grain crops and make harvest more difficult. Quackgrass prefers wet soils. Eliminate wet areas by digging drainage channels or installing tile drainage. Use a spring-tooth harrow for secondary tillage on a hot, windy day to pull a lot of the roots to the surface, where they dry out and die. Using a disc harrow, especially when the soil is damp, chops up the rhizomes and can allow each piece to grow into a new plant.

Growing a crop of buckwheat was an old way to clean up a field that had been grown over with quackgrass. It is planted late, allowing the field to be harrowed several times when the soil is dry to pull up roots. The buckwheat then smothers the remaining quackgrass, leaving the field free of weeds.

Species with a similar life cycle to the crop will tend to populate with that crop. For example, Canada thistle or sow thistle forms dense patches that can completely crowd out the crop. They grow in areas where drainage is slow and where soil has become compacted during field operations. Deep-rooted thistles spread when their root systems are not disturbed and persist in row crops that are not deep-tilled. Research done in the 1920s in Germany revealed that the thistles form an association with an anaerobic fungus that inhabits their root zone. Areas where they grow may have compacted subsoil that stays anaerobic for much of the year. In Europe thistles were traditionally controlled by mowing the patch just as they began to bloom. This prevents seed formation, but more importantly starves the roots by cutting off the tops just when root reserves are at their lowest. When left alone, thistles will eventually loosen the soil and relieve the compaction enough to alter the conditions that allowed them to grow so well. They in effect work themselves out of a job.

Unfortunately, if the soil compaction-causing practices that created the problem continue, the thistles will thrive and spread. Not tilling the soil when it's too wet can help correct the problem, but that may not be possible if the weather doesn't cooperate. Subsurface drainage can help remove some wet spots. Deep tilling after harvest when the soil is dry can help break up the compaction. Moldboard plowing destroys the large roots near the surface and reduces the vigor of the thistles and disrupts their life cycle, especially in the spring after they have started growing.

The most effective strategy to eliminate thistles combines improving the soil conditions and changing the rotation so that they can't complete their life cycle. With a long rotation of hay, the constant mowing prevents them from replenishing their roots and helps loosen the soil. In regions where the season is long enough, winter barley is harvested before the thistles finish blooming and a second crop such as beans or buckwheat can be grown in the same year. That cuts the thistles off before they can rebuild reserves, and the tillage for planting the second crop falls at a time when the thistles need to set seed and rebuild root reserves. Following that with a year of row crops can eradicate a tough stand of weeds.

Biennial Weeds

Several fall biennial species of weeds have exactly the same life cycle as winter grains. Cheat, corn cockle, vetch, corn chamomile, winter canola, and others can build up in winter grain seed when farmers plant bin-run seed without cleaning it or checking it for purity. Most such species have a very short seed life span in the soil, so that when they are cleaned out of grain seed, they will disappear after a few years. Planting clean seed is very important for controlling many weeds.

Allelopathic Weed Suppression and Organic No-Till

When I first started growing grains, I grew out a handful of black winter emmer in my garden bed. Midspring I went to the grain beds to weed and was pleased to see that the emmer bed was clean and weed-free. Later that day I thanked CR for weeding my emmer. He looked at me quizzically and said, "Huh? I did not weed any of the grains yet." Who weeded my emmer?

Did a forest elf sneak in at midnight to help? This is how I first learned about the astounding natural herbicide process called allelopathy.

Klaas Martens, our einkorn grower, reports on his 2015 einkorn harvest:

Eli,

We just finished harvesting and planting last Friday. There are still 5 acres of cover crops to sow but that doesn't worry me much. Everything else is done now! We have been at several conferences already and have lots more scheduled this winter. The older we get the busier we become.

Your einkorn is on two wagons in our shed. The straw was exceptionally nice. Our Mennonite neighbor baled it and was quite pleased with it. He takes straw from us and sends it back transformed into manure after he "uses" it. We get back all of the minerals plus the added nutrients.

No Weeds! The einkorn fields surprised me after harvest. While it was not competitive early on and the soil did not fill out with einkorn until late in spring, there were no weeds in those fields this summer or fall! Nothing! Einkorn has an amazingly strong allelopathy. The effective late season weed suppression could be very important in no-till crops seeded into rolled down einkorn.

If that turns out to work as we hope, this opens up a totally different use for einkorn seed that is still in the hull. Even at a high price, the small seeds and low seeding rate make einkorn an economical cover for crimped no-till soybeans. It seems that your einkorn still has a few new surprises left to show us! We will be doing more observation trials to try to measure its weed suppressing power.

I notice big differences in weed suppression between grain species and even between different varieties of the same species. Spelt and wheat fields have strong weed growth by late summer even with good cover crops. With the intense interest in no-till organic using crimped grain cover crops, this is important information – especially if my obser-vations can be repeated and documented. The other big questions will be how it responds to the crimper and at which growth stage it can be terminated. If it can give strong weed suppression even with slow early growth, with its great drought hardiness it can thrive in dry spring seasons when water shortages reduce soybean yields in our rotation system. Your winter einkorn could be uniquely valuable in organic no-till systems!

Have a good day!

Klaas

DISEASE MANAGEMENT

Soil health is the single most effective strategy to prevent soilborne dis-eases.[33] Any practice that stimulates healthy biological activity in the soil helps to suppress disease and cycles minerals to make the following crop stronger and more resistant to infection. A healthy, organically managed soil is a disease-suppressive soil. Organic strategies to control diseases inte-grate several practices: crop rotation, cool-managed disease-suppressive compost, on-farm selection of resistant plants, residue incorporation, staggering flowering times, and seed treatments. Abundant use of compost managed at low temperatures, combined with a three-year cover cropping rotation, encourages disease-suppressive microbial communities that are natural competition against disease organisms.

Fields that will be planted to winter small grains should have shallow till-age after the harvest of the previous crop. Discing, skim plowing, or other shallow tillage as soon as possible after the previous crop is off stimulates lost grains and weed seed to sprout and mixes crop residues with soil to begin the decomposition process. Shallow incorporation of crop residues helps eliminate disease inocula before the soil is prepared for the next crop. Volunteer grain is also eliminated by this practice, removing another important source of disease. By destroying volunteer grains, shallow tillage helps control late-season insects, which are vectors for virus diseases in winter grains.

Fusarium Prevention

Fusarium is the greatest disease pressure in the Northeast due to our typical rainy weather, susceptible cultivars, and crop residues left on the soil. Fusarium fungi normally do not attack the live plant. Their role is to aid in the breakdown of dead matter. When moisture is high at flowering and grain filling, tiny spores of fusarium infect the spike. Fusarium fungi produce mycotoxins and aflatoxins in the grain. Mycotoxins are produced by certain types of fungi that grow on maturing grains. Wet, rainy, warm, and humid weather from flowering time onward promotes infection of cereals by mycotoxin-causing fungi. Infection with mycotoxins is common on kernels damaged by insects, birds, mites, and moist weather.

Grains contaminated with fusarium vomitoxin have reduced quality, flavor, and yield. The presence of fusarium toxins may not be noticed until after there is an adverse reaction from eating the contaminated grain, so testing is essential for commercial grain crops. Mycotoxins make the grain unpalatable and can even be toxic. Flour mills will reject grain with more than two parts per million. Growers recognize fusarium in the field by the bleached out, white spikes that may ooze pinkish-orange spores. Infected heads have shriveled, lighter-weight seeds. *Aspergillus*, also a common soil fungus, causes stored grains to heat, decay, and produce aflatoxins.

Strategies to control fusarium include:

Crop rotation. Fusarium can survive on crop residues of previously infected crops. A three-year rotation between cereal crops will ensure complete decomposition of infected residues. Growing a nonhost crop for even a single year between cereals helps to reduce fusarium potential. Incorporate cover crops and diverse crops that are not hosts for fusarium and that build soil life.

Disease suppressive soil. Manage compost at lower temperatures with minimal turning over longer periods to encourage earthworm activity. Earthworm castings contain complex microbial communities that suppress fungal pathogens, and help decontaminate infected grain stalks that host fusarium. Liberal applications of living compost prior to seeding suppress fusarium.

Residue management. Avoid overwintering grain residues that can host fusarium. Contamination is common when wheat follows corn, especially if residues remain on the soil surface. Chopping and tilling residues enhances decomposition and decreases contamination.

Stagger planting dates. Or plant varieties with different days to maturity to reduce the risk of the entire crop being infected during flowering or grain fill, the vulnerable periods for infection.

Remove infected kernels postharvest. Since fusarium-infected kernels are lighter than healthy kernels, they can be removed using a gravity table or air column to separate out the lightweight kernels. A homemade gravity table can be fashioned simply by placing grain in a handheld box and shaking it in small vigorous motions so the lighter seeds are vibrated to the lower section to be removed.

Apply fungicides. Jack Lazor, of Butterworks Farm in Vermont, observes: "Spraying the biodynamic silica prep on wheats reduces fusarium." Silica is used to suppress fungal diseases, to stimulate leaf growth, and to enhance ripening, but it may cause burning if the weather is very dry. Contact the Josephine Porter Institute (jpibiodynamics.org) to order.

Grow and breed fusarium-resistant varieties. Growing resistant varieties is the most effective solution for organic growers. Where can we find fusarium-resistant wheats? Progress in breeding resistant wheat has been slow. To enhance fusarium resistance in your favorite varieties, harvest the healthiest wheat spikes on disease-free plants by hand. Rogue out infected plants and discard. This method works best with genetically diverse heritage varieties.

Klaas and Mary-Howell Martens give the following advice to large-scale grain growers to control fusarium mycotoxins:

1. Always use cleaned, high-quality seed that is not carrying seedborne diseases. Even if the seedborne diseases themselves do not produce mycotoxins, they can weaken the plant and damage the grain, which then can lead to infection by mycotoxin-causing fungi.
2. Harvest at maturity. To minimize grain damage, store when moisture content reaches 12 to 14 percent.
3. Adjust the harvesting equipment for maximum cleaning with higher blower speeds to remove small, shriveled kernels that may be diseased. Grain contaminated with foreign material is much more likely to develop mold and insect problems in storage.

4. Dry all grain to at least 14 percent moisture as rapidly as possible, at least within twenty-four to forty-eight hours after harvest. Safe, long-term storage can be achieved at a moisture level of 13 percent or somewhat below.

5. Cool the grain after drying and maintain dry storage conditions. It may help to clean the dried grain to further remove diseased kernels and weed seeds, but do not feed the screenings to animals.

6. Thoroughly clean the grain and bins before storage to remove crop debris, chaff, and broken kernels. Mold-infected kernels are easily broken. Damaged kernels are more likely to be contaminated.

7. Store in water-, insect-, and rodent-proof structures. Continue periodic aeration and probing for "hot spots" at intervals of one to four weeks throughout the storage period.[34]

Ben Gleason, an organic grain miller and grower in Bridport, Vermont, advises to sift out the bran in the flour to decrease the fusarium levels, since the outer surface of the grain carries the fusarium.

Using Biodynamic Sprays to Control Fusarium

In equisetum the cosmic is present, so to speak, in very great excess, yet in such a way that it does not go upward and reveal itself in the flower but reveals its presence in the growth of the lower parts.[35]

— RUDOLF STEINER

Having used biodynamic preparations[36] for the last seven years, Philip Lyvers, a biodynamic farmer, says, "Before using biodynamic preparations there were problems controlling fusarium diseases in the corn and wheat." To prevent fungus problems, Philip combines BD #508 (*Equisetum arvense*), BD #500 (Horn Manure), and Biodynamic Barrel Compost. He applies them to the soil in the spring before planting corn and on wheat before it heads at about 18 to 24 inches high. In the fall he applies them to the same acres before planting winter cover crops of crimson clover or oats. The BD #500 and BD #508 sprays enhance the decomposition of plant residue left in the field from his no-till method. This form of composting in the field is one of three composting techniques used by biodynamic farmers. According to Philip Lyvers:

Biodynamic Preparations

The Biodynamic Preparations (BD Preps) are herbal treatments to enhance the fertility, nutrient absorption, and health of soil and plants based on the teachings of Rudolf Steiner. The six BD compost preparations are made from yarrow, chamomile, nettle, oak bark, dandelion, and valerian. I have conducted scientific trials with and without the silica prep. There was less fusarium on the wheatheads with silica!! Try it yourself.

500: Cow horn manure is used to enhance soil fertility and renew degraded soils. A cow horn is filled with cow dung, and buried in soil over winter. In spring it is stirred vigorously into water by making vortexes in opposite directions to draw in energy, then sprayed on the soil.

501: Cow horn silica is used to bring in sun forces to enhance photosynthesis and to prevent fungal diseases, such as fusarium. Finely ground quartz crystals are placed in a cow horn and buried in soil over the summer, then stirred in water like preparation 500. The liquid is sprayed as a fine mist on plants in the morning.

Preparations for Compost

502: Yarrow flowers fermented in the bladder of a stag.

503: Chamomile flowers fermented in a cow intestine.

504: Stinging nettle fermented in the soil.

505: Oak bark fermented in the skull of a domestic animal.

506: Dandelion flowers fermented in a cow's intestines.

507: Juice pressed from valerian flowers.

508: Boiled horsetail, applied in excessively wet years to prevent fungal diseases.

The traditional rotation when cropping wheat is to plant the field into soybeans when the wheat is harvested. Instead of this we are using red and crimson clover to follow the wheat. We spray the preps and let it stay on the fallow clover until the following spring when we plant corn. [Rudolf] Steiner says that clover-nitrogen needs to be produced in November for use by the crop in the following year. In years with high rainfall there are people who used a fungicide but still had fusarium problems. The people who did not use fungicide had a bad-quality crop. I didn't use the fungicide, only the BD preps, and my wheat was good quality with little damage from fusarium. It worked. My flour is better than anything I can buy.[37]

Disease Control through Grain Storage

Standing out in your field, surrounded by those amber waves of grain, it is hard to imagine that the value of this beautiful crop could be spoiled by mishandling during harvest, transport, and storage. Growing the crop is only half the battle, but it is the part that farmers tend to focus most on. Our conventional farming experiences have mistakenly taught us that quantity, not quality, is all that really matters.[38]

— KLAAS AND MARY-HOWELL MARTENS

SANITATION

A key aspect of disease control in grains begins with sanitation in storage. The annual cleaning out of everything that can support yeast will also remove potential sources of fungal spores and toxins. Everything that prevents the growth of yeast also prevents the growth of disease-causing fungi in stored grains. Thoroughly clean all grain storage bins, repair leaks, and remove any possible sources of condensation in bins and other grain contact surfaces. Remove all stored grain. Thoroughly empty and clean the bins to help prevent infection and contamination of grains. Harvest equipment, trucks, and augers can also harbor infected grain. Preventing insect damage prevents fungal contamination. Insect feeding damages grains, opening them up to infection, and it creates areas in bins that attract moisture, further increasing the risk of contamination.

Klaas Martens advises:

I grow grain for traditional Shmurah Matzah,[39] the "Watched Matzah" baked by Hasidim in Brooklyn for the Passover celebration.

Generations of indigenous knowledge are embedded in these excellent traditional practices to prevent moisture from coming in contact with the grains — especially during storage. This controls the spread of critical diseases. I recommend to follow the Ancient Israeli guidelines to grow and store grains as explained in detail in the Talmud — for all end-use purposes of flour.[40]

Only clean, dry grain should be put into storage. A low moisture level of the grain going into storage is the key to successful storage. For poor-quality grain produced under drought or moisture stress, or with a large amount of damaged kernels, the recommended maximum moisture levels should be 1 percent lower. Note that this percentage refers to the highest moisture in the bin, not the average moisture. As little as 0.5 percent moisture can mean the difference between safe storage and a damaging invasion by storage fungi. Check your bins. Bins should hold the grain without leaks of rain or snow from outside, without access for rodents and birds, without contact with soil, and with sufficient headroom to permit sampling, regular inspection, and ventilation.

Be sure that headspace air is well ventilated. If you pack a bin to the roof, water condensation where the grain touches the roof may result in mold, which can spread downward through the grain. Warm, moist headspace air can activate mold growth, causing grain to crust and seal over. Mold can spread over kernel surfaces, resulting in caking near the surface in about three weeks. It will also produce carbon dioxide along with water and heat, raising the humidity and causing further mold growth. Crusting and caking is not just a grain-quality problem: the chunks can get stuck in augers, making unloading the bin really difficult.

Check it regularly. Grain may become moldy when left undisturbed for several months in a bin. A regular, monthly inspection routine throughout the grain storage time is important, especially during the summer and early fall months, when grain temperatures are optimal for rapid insect and mold development. During warm weather, insect infestations generally begin near the grain surface, usually at 6 to 12 inches depth, and especially at the point of entry where dust, broken kernels, and chaff accumulated during filling. Walk in the grain bin. Your feet should sink in slightly. If the grain feels hard, this could indicate caking. Smell the grain for mold, look for insects, and stick your hand down at least 10 inches to check for heating and caking.

Take out a load about a month after filling the bin. This breaks up any moldy grain that might have formed at the surface.

GRAIN MITES

Diatomaceous earth is a nontoxic, food-grade organic solution to control mites. It is a white powder that comes from fossilized freshwater algae. When mixed into the grains it kills the mites and keeps them away. It is an effective alternative to the toxic fumigation process that nonorganic operations use to eliminate common mite issues. If mites are in your grain storage bin, these little creatures eat the kernels, which kills the seed. The entire lot will be destroyed. The diatomaceous earth is abrasive to the mites' bodies and dries them out. Diatomaceous earth is high in silica. Users beware: Breathing it is dangerous and may cause lung problems. Touching it may severely dehydrate your skin.

CONTROL GRAIN MOTHS BY FREEZING

The presence of light webbing or tiny holes in the grains indicates infestation by grain moths. Infested grains should be discarded or treated by freezing. Freezing the grain kills grain moths. Place infested seeds in a deep

In July the tall red-eared varieties of wheat with strong stalks went through a whole palette of color changes. At first the ears were a pale olive green, then they became more brown, while at the same time the stalks began to glow a salmon pink, as if they were painted with a fluorescent marker pen. It was not so much the color as the glow that appeared with it. The stalk was permeated with light as if it were illuminated from behind or within. The effect was strong. It conveyed an impression of summer warmth and sunshine which however, was not without a touch of melancholy. It was not the beauty of blossoming, but rather an autumn beauty of release, of letting go. Gradually the glow disappeared and the dense, reddish brown straw color remained.

Christine Karutz[41]

freeze for at least a week. Effectiveness of the freezing treatment is improved by alternating freezing treatments with thawing to room temperatures so that any larvae that survive will subsequently be frozen.

Harvest Arts for Small-Scale Growers

Farmers would do well to bear in mind that wheat is greatly injured in quality and quantity by allowing it to get too ripe. The farmer will be deprived of his reward. When the wheat is left to mature to an advanced condition the husk is thickened and the yield certainly lessens. Wheat should be cut the moment the stem changes in color from green to yellow-brown, and when the grain, when pressed between the finger and thumb, no longer gives out a milk juice, now being in the hard dough stage. At this moment the wheat should be reaped as speedily as possible. Not only will the grain be fine, but the straw will be sweet fodder for animals, richer in nutriments. The theory that explains that the sugar is changed to starch in the grain, but if permitted to remain until fully ripe, another change takes place. The starch is gradually converted into woody fiber.[42]

— S. EDWARDS TODD

When shall we harvest wheat? Deciding when to harvest and how to handle the grain after harvest are critical. Harvest time depends on whether your end use is for flour or for seed saving. The big picture is more important than knowing average harvest dates for a specific area. While we can draw charts with footnotes to determine when the crop should be harvested, it is more effective to understand how to determine when the crop is mature and base your decision of when to harvest on combined factors that include end use, harvest weather, equipment availability, drying capacity, and available storage facilities.

HARVESTING FOR FLOUR

For flour, cut the grain about two weeks before it is fully ripe. At this stage, the grain will weigh most, giving the largest proportion of flour and the least bran. If the crop is left uncut, a thicker bran will form and a portion of the starch of the grain will change into woodlike fiber. The quality of flour is decreased and the weight of bran increased.

When the straw immediately under the head of grain turns from a greenish to an orange hue 4 or 5 inches in length, it is time to cut the grain for flour. The kernels have passed out of the milky state but can be easily crushed between the thumbnails. At this time, some of the leaves on the lower portion of the stem may be dry and brown, but still, the upper stem remains vigorous for a few days. Do not delay the harvest. Harvesting wheat while it is still slightly green and drying in the barn or stooking outdoors to dry-cure the grains produces the highest-quality wheat flour.

After wheat begins to dry in the field, rain will decrease its quality and test weight, causing fusarium and sprouting damage. Quality decreases daily if wheat stays in the field after maturity. The wheat will have the highest test weight and best grain quality if it is immediately harvested after it is mature and dry. If wheat stands in the field after maturity and is rewetted by rain or heavy dew, the grain shrivels, test weight is reduced, sprouting can occur, and field loss is increased.

HARVESTING FOR SEED

If you are harvesting for seed to replant, allow the plant to reach full maturity before cutting. At full maturity the grain has received the full life force from the stalk and will have a higher germination rate as long as there is not sprouting or moisture damage. Harvest time depends on the weather. A damp, cool spring will give a late harvest, whereas a warm, dry spring will ripen the wheat up to a month sooner.

Finally, the crop yield should be calculated and recorded. Use the observations made during the year together with your harvest records to find which factors were yield-limiting, and to evaluate whether the stand density was adequate, too low, or too high. Carefully evaluating the previous year's management decisions based on the results helps build knowledge to make better-informed decisions on future crops.

Traditionally wheat was cut about two weeks before full maturity, then cured in the field in stooks. Before the invention of the combine, farmers scythed at the hard dough stage, when the stalk was still slightly green in the upper third, then tied the sheaves in bundles that were placed in shocks or stooks to dry slowly. The grain was later threshed when fully dry. Early harvest and stooking allow farmers to harvest quickly before an impending rainfall, protecting the mature grain from damaging moisture. In nineteenth-century New England, when the combine and the reaper-binder were both available, farmers preferred the reaper-binder that bound into stooks, since the slow-curing process of stooking imparts higher quality and flavor to the grain. Harvesting with the combine works only when the grain is totally mature and dry, making harder bran. Stook-curing is well suited for harvesting high-value heritage landraces that mature at slightly different rates.

SCYTHING

Join the scything renaissance! Every farmer and gardener should have a scythe. It is a forgotten tool that is being rediscovered as an easy-to-use, affordable, and practical alternative to peak-oil machinery. Swinging the light European, a.k.a. Austrian, scythe is an invigorating, whole-body exercise similar to tai chi. The mower swings the blade back and forth, letting the weight of the scythe carry the motion in a comfortable rhythm. Lightweight European scythes are a zero-carbon method to cut meadows, weeds, and even suburban lawns. The English scythe used in colonial America is heavier and more difficult to swing. More muscle, less Zen. I bought a scythe from each of the five North American scythe companies, since I often have interns at harvest time. All the US scythes made in the European style that I tried are well made. Differences were slight. Swish swish!

CR Lawn reports, "The scythe is one of the two tools that I cannot do without on my farm. It gives me freedom to manage all of the grass to cut for mulch. It is part of my whole system. It is a great way to start the day early in the morning when the dew is still on the grass. My other favorite tool is the potato hook — the best weeding tool there is."

When harvesting grain with a scythe, you need to attach a cradle to move the wheat stalks neatly to one side so that they lie in the same direction for gathering. Without a cradle on your scythe, the wheat stalks fall every which way. A simple bow cradle can made by lashing a green branch to the scythe with string to push over the wheat stalks.

I use a sickle more than a scythe when I select rare wheats. The sickle allows me to select the exact ears that I want to harvest, whereas the scythe takes all. If you use a sickle, one with a serrated edge is the best.

THRESHING

In the United States, as the number of small-scale grain growers increases, there is a great unmet need for smaller threshing machines. In developing countries, small-scale grain growers use appropriate technology, spanning small-scale machinery, animal power to tread on the grain to loosen it from the seedhead, and flails. Flails use two heavy sticks bound together with a leather tie so that the shorter piece lies flat on the grain. I have threshed small amounts at good speed by placing the cut seedheads on an upside-down floor mat from a car, knobby side up, over a tarp or in a large container, then doing the wheat shuffle dance. Although common in Third World countries, few treadle threshers are commercially made in the United States. Small-scale US grain growers are devising ingenious machines, or importing them from abroad. Mark Fulford imports international small-scale threshers to meet this growing market.

Winnowing

Pouring out the threshed grain in front of a fan separates out the good heavy seed from the lighter. I use a commercial vacuum blower to blow off the chaff, then a clipper seed cleaner with vibrating screens to separate out the lighter seeds. The smaller seed and debris fall through the holes, and the fat seed is collected.

DEHULLING

Farmers who grow hulled grains must find an effective way to remove the hulls. Dehulling poses a similar problem to that of threshing. Most

Hulled Grains Make Better Beer

Another method for processing hulled grains is to malt, mash, and ferment them into beer. Instead of being a problem, the hulls aid in the malting process by filtering, which extracts the liquid beer after the mash is fermented. Whole barley with the hull is processed to make malt for beer. The hull is important in the malting process to prevent spoilage in the malt vats.

equipment is designed for high volumes and large fields. The machinery is expensive, and its cost must be spread over a large volume to be economical.

Pounding

Before hulled grains can be cooked or milled into flour, the hulls must be removed. In early times, hulls were separated from the kernels by pounding with a heavy wooden pestle or mallet. The hulls were then winnowed out. Pounding the grain by hand, as has been done for millennia, works well.

Flaming

Grain harvest celebrations, from the traditional Celtic Lughnasadh to the ancient biblical Shavuot as recorded in the Talmud, involved soaking the sheaf of hulled grain in water, then singeing it in fire to loosen the hulls. This tasty, smoked, charred grain is today known as *frikah* in Arabic and *kaule* in Hebrew. After the hulls are loosened from the kernels, they are separated from the grains by threshing and winnowing.

A Home Dehuller

Increase the space between the stones of a stone mill so that the hulls are abraded and rubbed off and so that kernels can pass through without being broken up too much. You can also use a burr mill that has had one of the burr wheels replaced by a hard rubber disc to rub the hulls from the grains. The loose hulls should be winnowed out of the mixture of grain and chaff afterward. Old flour mills that ground spelt in Europe were equipped with two pairs of mill stones. The first pair was located above the second. The bottom stone of the first pair turned while the top stone was stationary. The spacing between the two stones was set to just allow dehulled kernels of grain to pass between them. The mixture of kernels and hulls fell from the top pair of millstones into the second pair after falling through a column of air that blew out the loose hulls, leaving the grains to fall into the second pair of stones that then ground them into flour.

To modify a threshing mechanism for dehulling, the spaces in the concave between the bars need to be filled with flat steel strips to prevent the grain from falling through the concave. With the concave thus closed, the grain stays between the cylinder bars and the concave bars, where it is rubbed apart. A lot of the dehulling is accomplished by rubbing grain on grain when the grain can't fall through the concave. The advantage of using a combine or threshing machine is that the mixture of grain and chaff also passes through the cleaning sieves and is winnowed by the fan afterward.

Using any of the alternative dehulling methods results in leaving some degree of foreign matter and hulls in the grain. After grinding grain that contains excess hulls and other fibrous material, you need a screen or sift to remove the objectionable material. A flour sifter will work if nothing else is available.

Small grain growers may find it affordable to have the grain cleaned and dehulled at a commercial facility with cleaning and dehulling equipment.

A dehuller manual written by Nigel Tudor is available for a free download on the SARE website.[43] The well-illustrated instructions clearly explain how to fabricate a small-scale dehuller based on a German model.

DRYING THE GRAIN

Modern wheat harvesting methods use artificial forced drying that suppress flavor and quality.

— GLENN ROBERTS

The grain must be fully dry and cleaned of weeds and other foreign materials before milling. A low-tech way to dry small quantities of grains is to spread them on a wooden floor and to shovel or rake them every few hours until they are dry. A box fan blowing over the grain speeds the process. A drying box can be improvised by cutting a hole in the bottom of a wooden box and placing a screen mesh on it. Place a fan underneath. Fill it with grain, and let the slow aeration dry the grains. It's important to keep the grain stirred and loose. Wood helps draw moisture out and prevents condensation at the bottom of the pile.

Few farmers today are aware of the traditional slow-curing process that produced lighter, digestible bran with more richly flavored grain. Large-scale growers use grain-drying equipment to decrease kernel moisture. This maintains the quality of the grain while minimizing spoilage. The grains can be dried by inserting an aeration hose into perforated totes, using drying fans, and placing the grain on pallets to increase air circulation. Bins must be cleaned meticulously before filling them with grain. The grain moisture content can be tested by trying to dent the kernel with a fingernail till it is fully hardened and dry. The wheat is ready to store when it is very hard to crack it between your teeth.

SEED DORMANCY

A wheat seed is dormant when it will not germinate even under good conditions. Seed dormancy right before harvest prevents premature sprouting. Dormancy increases with lower temperatures during grain fill. Postharvest dormancy generally dissipates by late September to early October. Hot or cold conditions promote germination better than seed stored at ambient air temperatures. If you plan to sow early or conduct a germination test, refrigerate seed for two days prior to planting or testing.

In black winter emmer, each spikelet has two kernels. One germinates immediately. The other lies dormant for a few weeks, then comes alive as a mechanism to protect against adverse weather conditions.

Einkorn

Einkorn (*Triticum monococcum*), an ancient, almost-forgotten species of diploid wheat from the dawn of agriculture, is being rediscovered today as a power grain, higher in protein and minerals than modern wheat, yet safe for most gluten allergies. Why? Einkorn is not genetically related to modern wheat. All modern wheat evolved from wild emmer wheat (*T. dicoccoides*), except einkorn, which evolved independently from wild einkorn (*T. baeoticum*).

Einkorn's delicate, filigree glumes rise up from the grain head, imparting a feeling of calm beauty. It can be mistaken for a wild plant. Herein lies its value to peasants through the ages. Einkorn's wild resilience and strong symbiosis with the mycorrhizal fungi that scavenge soil nutrients enables it to thrive like a weed in drought and harsh conditions where modern wheats fail. Einkorn has been cherished by villagers through the ages for its rich flavor, digestibility, and high nutrition. It contains more antioxidants, protein, and minerals than modern wheat. It is higher in protein than quinoa. In fact, einkorn is higher in protein than any other small grain. Einkorn's close association with mycorrhizae enables it to reach out to a large area for nutrients. Its wildlike ability to translocate nutrients into the seed translates into an average of 30 percent higher protein; 15 percent less starch; 200 percent higher lutein; 50 percent greater manganese, riboflavin, and zinc; and about 20 percent higher magnesium, thiamin, niacin, iron, and vitamin B_6 than modern wheat.

Since einkorn evolved in lands with harsh conditions, only the most resilient survivors of adversity have come down to us today. Einkorn not only has an ability to scavenge more soil nutrients than modern wheat, but it even builds soil like a cover crop due to its strong association with mycorrhizae. It also carries durable resistances to the fearsome UG99 strain of rust that threatens to decimate world wheat crops and has a salt tolerance that enables it to thrive in saline soils.[44] High salt levels have degraded 20 percent of the world's farmland due to irrigation and evaporation, which cause salt buildup in the soil, posing a serious threat to world grain production as we face the high rainfall and temperatures of climate change. Australian scientists have incorporated einkorn's salt tolerance into durum wheat, boosting grain yields by 25 percent in saline soils.[45] Instead of merely using einkorn for breeding traits into modern wheat, why not use einkorn itself? This is equally true for the other neglected grains like *T. timopheevii* and *T. carthelicum*, which have higher resiliences, flavor, and nutrition than modern wheat.

TRADITIONAL EINKORN FARMING IN THE CARPATHIAN MOUNTAINS

Due to einkorn's rich flavor, vigor, and adaptability to the thin soil of stony mountain fields, traditional peasant farmers have continued to cultivate it in small, remote villages in Europe.[46] Free-threshing bread wheat (*T. aestivum*) was cultivated in more fertile valleys; however, in the hilly regions, free-threshing wheat, with its higher need for fertility, did not play a significant role before the Middle Ages. The peasant farmers who still grow einkorn use hand tools such as scythes, sickles, spades, and hoes with animal traction or human labor.

In the villages where black winter emmer and einkorn are still grown, einkorn is used for making bread and porridge and is highly valued as a poultry feed that produces high-quality eggs with bright yellow yolks. A field study of peasant farmers who grow einkorn conducted by Dr. Mariá Hajnalová and Dr. Dagmar Dreslerová documented traditional practices that are a window into age-old grain farming methods.[47]

The Harvesters by Pieter Bruegel the Elder, 1565. Five-foot-tall wheat in 1565. Heritage wheats are tall with deep root systems that evolved in organic soils over millennia. Courtesy of the Metropolitan Museum of Art, New York City.

One-foot-tall wheat in 2000. Dwarfed, woody modern wheat hybrids are bred dependent on agrochemicals to survive, and have about 500 percent less leaf surface area than heritage wheat. Courtesy of Martin Pettitt.

Em Ha'Hitah, wild emmer wheat, the mother of all cultivated wheat.

The author, Eli Rogosa, with Klaas Martens, in Penn Yan, New York. Klaas, a gifted grain grower, has multiplied Eli's einkorn breeding population into a thriving field of abundance.

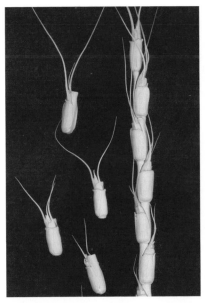

Goatgrass (*Aegilops tauschii*). Courtesy of USDA Agricultural Research Services, Germplasm Resources Information Network (GRIN).

Miracle; a branched wheat sub-type of Rivet/Cone (*Triticum turgidum* subsp. *turgidum*). Courtesy of GRIN.

Einkorn (*Triticum monococcum* subsp. *monococcum*). Courtesy of GRIN.

Cultivated emmer (*Triticum turgidum* subsp. *dicoccom*). Courtesy of GRIN.

Wild emmer (*Triticum turgidum* subsp. *dicoccoides*). Courtesy of GRIN.

Durum (*Triticum turgidum* subsp. *durum*). Courtesy of GRIN.

Vavilovii (*Triticum aestivum* subsp. *vavilovii*). Courtesy of GRIN.

Barley (*Hordeum vulgare* subsp. *vulgare*). Courtesy of GRIN.

Khorasan, a.k.a. Kamut (*Triticum turgidum* subsp. *turanicum*). Courtesy of GRIN.

Persian (*Triticum turgidum* subsp. *carthlicum*). Courtesy of GRIN.

Poulard, or rivet (*Triticum turgidum* subsp. *turgidum*). Courtesy of GRIN.

Zanduri (*Triticum timopheevii* subsp. *timopheevii*). Courtesy of GRIN.

Bread Wheat (*Triticum aestivum* subsp. *aestivum*). Courtesy of GRIN.

Indian (*Triticum aestivum* subsp. *sphaero-coccum*). Courtesy of GRIN.

Macha (*Triticum aestivum* subsp. *macha*). Courtesy of GRIN.

Spelt (*Triticum aestivum* subsp. *spelta*). Courtesy of GRIN.

Traditionally, heritage wheat was cut about two weeks before full maturity, then dried down in the field in stooks (piles of sheaves) to cure and dry for a softer, more digestible bran and richer flavor. Courtesy of Mark Robinson.

A Mesopotamian grain goddess receiving gifts from a farmer-god and a human worshipper. Drawn from a cylinder seal, circa 2350 to 2150 BCE. Drawing copyright © Stéphane Beaulieu.

"The Wheat Itself." This fertility figurine depicts two lovers in a sacred embrace. It was found in the Wadi Haritoun cave in Tekoa, Israel, fashioned approximately 11,000 years ago by the early Natufian farmers who domesticated wheat. It reflects the sacred embrace of the self-pollinating male and female aspects of the wheat itself. Photo courtesy of Geni, Wikimedia Commons. Figurine from the British Museum, London.

Madonna of the Ears, Tyrol 1495, attributed to Rueland Frueauf the Elder. Courtesy of the Salzburg Museum.

Ethiopian wheat. Ethiopia is a center of wheat biodiversity, with treasures of diverse purple wheats.

CROPPING SYSTEM

The crop rotation consists of a winter crop followed by a root crop and then a spring cover-fodder crop of either alfalfa or clover, or vegetables. Fallowing with a forage crop such as alfalfa or clover is part of the crop rotation. The length of the fallow can be anywhere from one to three to five years or more. The fields were close to the village, divided into small plots from generations of splitting the inheritance among the children.

TILLAGE

The fields were either plowed by oxen or horses or hand-cultivated with a hoe and a rake. One farmer first burned the weeds and stubble, raked the burned surface stubble, and then made furrows by hoe and covered the seed grain by raking. In such a way, a plot could be tilled up and planted by an elderly woman farmer in one day. The farmers felt that hand cultivation was time consuming, hard work, but its results were very satisfactory.

FERTILITY MANAGEMENT

The rotation system consisted of manuring, followed by a vegetable or legume crop, then einkorn, then fallowing with forage crops such as alfalfa that also serve as a cover crop with animal grazing. Farmyard manure was traditionally plowed into the soil for a spring-sown vegetable crop that would precede the einkorn. Manure was not always applied to einkorn cultivated for straw, as the stalks become too tall with a tendency to lodge. Middle Eastern farmers that I interviewed stated that manuring cereals caused abundant straw growth, which led to lower yields. Farmers generally felt that one cow produced enough manure for 1 hectare (2.47 acres) of arable land. Plots of einkorn in Transylvania were on fields where manure was applied to the crop preceding einkorn. The high value of legumes as a cover crop was appreciated.

SEED SOWING AND SAVING

Sowing of einkorn was done by hand, either by broadcasting or planting in furrows. Hand cultivation could result in a harvest at least three times as abundant as when a plow was used. Most of the einkorn landraces were facultative, producing grain equally well when sown in autumn or in spring. Farmers do not import seed from elsewhere and save their own seed for sowing year after year. Similarly, they only plant local bread wheat or rye seed. The seed for sowing was selected from the best part of the harvest, from plants with the largest ears. The largest grains were separated out by hand sieving and saved for seed. Such practices help farmers maintain a good seed stock.

Winter or Spring?

All wild wheats are winter types. The first domesticated wheats had, as their wild progenitors, a winter growth habit. Spring growth evolved independently through spontaneous mutation and adaptation to colder climates. Spring types were selected for cultivation in regions where winter was too severe. The majority of the farmers preferred to sow einkorn in the fall, feeling that this brought a better crop both in grain and straw quality. If the sowing could not be done by late fall due to wet soils, early snow cover or lack of time, it was postponed until the spring.

To separate the grain from the husks, the einkorn seedheads are parched by fire, then pounded in mortars with a wooden paddle with repetitive winnowing and sieving. In ancient biblical times the seedheads were soaked in water right before parching over an open flame to loosen the hulls from the grain. Grain for consumption was dried in ovens to prevent infestation by insects. Einkorn flour was often mixed with potatoes or rye in bread.

CHAPTER 4

Journey of the Sheaves: Grain Folk Traditions

W hat we believe about our past shapes our beliefs of who we are as human beings and as farmers, our understanding of how food crops evolved, and our potential to coevolve with plants today. Just as the seeds are almost lost, the practices that evolved the unique qualities of landrace wheat have been forgotten. Plants sense, touch, hear, smell, communicate, and respond. Plants are intelligent beings that exhibit problem solving, planning, and defensive behavior. Plants actively search for nutrients and for sunlight. They move their leaves and stems in response to light and change the direction in which they grow in response to available nutrients. Plants such as the mimosa close their leaves when touched. Plants even respond to noise or music with changes in growth responses.

CR Lawn, founder of Fedco Seeds (fedcoseeds.com) and master farmer, sings constantly in his fields, everything from Gregorian chants and rounds to old Beatles songs. He explains:

Studies have shown that plants respond positively to beautiful classical music, and negatively to dissonance. Whether my singing falls in the former category or the latter I am unable to judge, but I would like to think the former. I strive to feel exalted and calm, but never angry in the garden, and hope my singing reflects that. Plants love attention. They respond to any positive attention, the more the better. It is good to thank them, to work near them and to apologize to them if you upset them with unnecessary clumsiness. They are highly sentient beings, smarter than we give them credit for, and in a deep dance of co-evolution with us. They need us just as we need them.[1]

Marija Gimbutas, a Harvard anthropologist, spent years collecting folk songs in the villages of eastern Europe in the 1930s. She observed:

After many years collecting songs from the farm women in the villages, I started to understand the ancient song, what the song was in the really beginning. People did everything singing. The song traversed the earth. Singing was at the heart of farm life from planting to harvest. The women working hard were happy because they had the song with them. Their belief system was expressed in what they sang in the songs.[2]

Just as modern breeders design today's farm animals to be stupid so that they are easier to manage and exploit, we have dumbed down our food crops. The first time I met heritage sheep, I was amazed at their intelligence and responsiveness. All the sheep I had ever met before were modern sheep, docile and mindless. No one was home. Whereas the heritage sheep looked me in the eyes, approached me thoughtfully, and asked who I was, as a dog would.

Walk in a modern wheat field. Harvest a sheaf of modern wheat. It is hard, woody, stiff, and unpleasant to touch. Nobody's home. Walk in a field of heritage wheat. It awakens a feeling of exhilaration and awe. The plants call to you with warmth and intelligence. Yes, intelligence — and certain varieties are more intelligent than others. They interact with me more and respond more quickly to changes in the environment.

Agriculture as a Divine Gift

In the mythology of almost all civilizations, the knowledge of farming is a gift from the gods. Each culture teaches a unique story with local sacred beings, but the divine nature of farming echoes throughout. In the Mediterranean, the source was the goddess Asherah in Canaanite and early Israeli cosmology. In biblical Israel, tradition teaches that Noah gave the people the plow. Isis in Egypt, Demeter in Greece, and Ceres in Rome taught the arts of farming. In China, it was the ox-headed god Shennong. In Mexico, it was Quetzalcoatl disguised as an animal. In Peru, Viracocha, the Inca sent by the Sun Father, taught the people how to farm.

It was Demeter who taught Triptolemus whose name means "he who pounds the husks,"[3]

to yoke oxen and to till the soil and gave him the first grains to sow. In the rich plains about Eleusis he reaped the first harvest of grain ever grown, and there, too, he built the earliest threshing floor. In a cart given him by Demeter and drawn by winged dragons he flew from land to land scattering seed for the use of men.[4]

The Royal Commentaries of the Inca recounts:

Know then that, at one time, all the land you see was nothing but mountains and desolate cliffs. The people lived like wild beasts, with neither order nor religion, neither villages nor houses, neither fields nor clothing, for they had no knowledge of either wool or cotton. Brought together haphazardly in groups of two or three, they lived in grottoes and caves like wild game, fed upon grass and roots, wild fruits, and even human flesh. They covered their nakedness with the bark and leaves of trees, or with the skins of animals. Some even went unclothed.

Seeing the condition they were in, our father the Sun was ashamed for them. He decided to send one of his sons and one of his daughters from heaven to earth that they might teach men to adore him as their god; to obey his laws as every reasonable creature must do; to build houses and assemble together in villages; to till the soil, sow the seed, raise animals, and enjoy the fruits of their labors like human beings. The Inca King and Queen arrived from heaven and were given a sign by which they would know where to establish a capital city. The place was located and they set out to teach the savages "how to live, how to clothe and feed themselves like men, instead of animals." While peopling the city, our Inca taught male Indians the tasks that were theirs, such as how to select seeds, till the soil, make hoes, and irrigate their fields by means of canals that connected natural streams, and even how to make these same shoes that we wear today. The Queen taught the women how to spin and weave wool and cotton, and how to make clothing.[5]

In Chinese mythology, the creator, Pangu, separated the heavens and the earth and created the sun, moon, and stars. Pangu created the plants and animals with twelve celestial sovereigns, one of whom was Shennong, a tree-being whose name means Divine Farmer. Shennong taught the people how to farm, how to use the hoe and the plow, and how to sacrifice at the harvest thanksgiving ceremony. Shennong had the body of a man and the head of an ox, and his element was fire. He discovered the healing uses of herbs that cured illness and made a five-stringed lute.[6]

BECOMING HUMAN IN MESOPOTAMIA: SACRED SEX, BREAD, AND BEER

The Epic of Gilgamesh is among the oldest written stories on Earth. It comes down to us from ancient Sumeria-Mesopotamia, written on twelve clay tablets in cuneiform script. It tells of the journey of the historic king Gilgamesh of Urak (modern-day Iraq) to find the secrets of life 4,500 years ago. The epic, the greatest surviving work of Mesopotamian literature, opens with the encounter of the god-man Gilgamesh, who abused his regal powers, forcing his people to build magnificent monuments and taking any woman in the empire that he desired. Enkidu, a wild, untamed human who ate grass and the milk of wild animals, wanted to test his strength against Gilgamesh, the demigod sovereign. Taking no chances, Gilgamesh sent a woman to Enkidu to learn of his strengths and weaknesses. Enkidu enjoyed a week with her, during which time she taught him the arts of civilization. Enkidu knew neither what bread was nor how one ate it. He had also not learned to drink beer.

This holy woman came to Enkidu, taming this wild man with sacred sex, bread, and beer, saying, "Eat the bread now, O Enkidu, as it belongs to life. Drink also this beer, as it is the custom of the land." Enkidu drank seven cups of beer, and his heart soared. His heart grew light, his face glowed, and he sang out with joy. In this state of elevation, he washed himself and became a human being.

Ancient beers not only lightened the heart but were brewed with herbs and contained natural antibiotics carefully brewed to create a healing drink.[7]

Hymn to Ninkasi

Recipe for Sumerian beer inscribed on a clay tablet from 1800 BCE.[8]

Borne of the flowing water,
Tenderly cared for by the Ninhursag,
. . . .

Having founded your town by the sacred lake,
She finished its great walls for you,
Ninkasi, having founded your town by the sacred lake,
She finished its walls for you,

Your father is Enki, Lord Nidimmud,
Your mother is Ninti, the queen of the sacred lake.
. . . .

You are the one who handles the dough [and] with a big shovel,
Mixing in a pit, the bappir with sweet aromatics,
Ninkasi, you are the one who handles the dough [and] with a
    big shovel,
Mixing in a pit, the bappir with [date]-honey,

You are the one who bakes the bappir in the big oven,
Puts in order the piles of hulled grains,
. . . .

You are the one who waters the malt set on the ground,
The noble dogs keep away even the potentates,
Ninkasi, you are the one who waters the malt set on the ground,
The noble dogs keep away even the potentates,

You are the one who soaks the malt in a jar,
The waves rise, the waves fall.

Ninkasi, you are the one who soaks the malt in a jar,
The waves rise, the waves fall.

You are the one who spreads the cooked mash on large reed
    mats,
Coolness overcomes,
. . . .

You are the one who holds with both hands the great sweet
    wort,
Brewing [it] with honey [and] wine
(You the sweet wort to the vessel)
Ninkasi, (. . .)(You the sweet wort to the vessel)

The filtering vat, which makes a pleasant sound,
You place appropriately on a large collector vat.
Ninkasi, the filtering vat, which makes a pleasant sound,
You place appropriately on a large collector vat.

When you pour out the filtered beer of the collector vat,
It is [like] the onrush of Tigris and Euphrates.
Ninkasi, you are the one who pours out the filtered beer of the
    collector vat,
It is [like] the onrush of Tigris and Euphrates.

MYSTERY OF GÖBEKLI TEPE

Some 250,000 years ago, Europe was populated by Neanderthals. Their matriarchal, moon-calendar culture may have been a primary source for ancient intuitive wisdom that survives today in our beliefs, myths, folklore, and religious practices. Neanderthals may have introduced an earth-based matriarchal, peaceful civilization to the Cro-Magnons who migrated up from Africa.[9] According to Stan Gooch in *Cities of Dreams: When Women Ruled the Earth*,[10] Neanderthals sanctified the number thirteen, associated with the lunar feminine moon cycles, and had a profound knowledge of nature, herbs, crystals, and minerals.[11] Neanderthals honored wild creatures such as the cave bear, the spider, and the serpent, with whom they shared their caves, and had complex burial rituals suggesting a belief in an afterlife, the very elements expressed in Göbekli Tepe, where einkorn and barley were first harvested.

This ancient megalith is shrouded in mystery. Over 12,000 years ago, possibly to a pre–Ice Age civilization, 7,000 years before the Mesopotamian civilization or Stonehenge, vast numbers of hunter-gatherers built circles upon circles of immense sacred structures in what is now modern Turkey. Thirteen years of careful digging has unearthed a mere 5 percent of the structures. Who built this complex? What was their purpose? The 30-acre temple complex of megalithic stone circles is surrounded by wild einkorn fields. Did the need to feed the people who built Göbekli Tepe lead to the gathering and cultivation of einkorn? If the lands around Göbekli Tepe were fertile enough to support the people to harvest and cook wild einkorn, agriculture may have arisen as a result of the need to feed the Göbekli Tepe people, not as a random decentralized discovery.[12]

Jack Harlan, the renowned plant anthropologist, challenged the common belief that hunter-gathers were driven by hunger to domesticate and cultivate plants. In Turkey he demonstrated how to gather two pounds of clean wild einkorn grain per hour using a stone-blade sickle. In a three-week period a family could gather enough grain to sustain themselves for a year.[13]

Grain Eaters at Ohalo, Israel

A great hoard of 16,000 carbonized grains was found preserved at Ohalo II, a submerged 23,000-year-old human settlement by the Sea of Galilee in Israel. The huge amount of grains found there confirms that wild emmer and barley were staple foods of the ancient hunter-gatherer-farmers.[14] At the center of the site are grinding stones and hearths, indicating that breads were baked.[15]

Old Europe

The Celts are among the earliest people who lived in Old Europe, spanning the Balkans to Scotland, from France to Turkey.[16] Celtic customs reflect pre-Christian traditions before Roman conquest. Marija Gimbutas describes the imagery of Old Europe as an expression of concepts of sacred cosmology within a mother-kinship structure. Gimbutas reports:

> The folk traditions from the Baltic to Caucasus Mountains are living memories to this day that reach back to the pre-Christian antiquity of early times still alive in Celtic culture. Folk traditions have faithfully preserved the ancient practices and beliefs which are closely related to the early traditions of the Indo-Iranians known as Zoroastrian, in the veneration of sun, water, fire and the life-processes of farming. These traditions reflect the peasants' perception of the world, their rich natural environment, sustaining a profound respect for the living land, forests, trees and flowers, and an intimate relationship with animals and birds. The legacy of European prehistory is rooted in the ancient religion, which is expressed in a cosmic, lyrical conception of the world at the center of their remarkable sustainability.[17]

The Old Europe farming societies that evolved had well-planned villages with soundly constructed homes, elegant sculptural and ceramic arts and traditions creating shared experiences that may have included storytelling, singing, and dancing, common in all indigenous societies, that harmonized the people within their specific ecosystems.

Who Were Europe's First Wheat Farmers?

The two centers of wheat's origin, the Levant and Caucasus Mountains, were the grain sources for Europe's early farmers.

Early Middle Eastern farmer-explorers brought wheat to Stone Age hunter-gatherers in Europe. Skeletons found in Europe's earliest post–Ice Age burial sites have DNA markers that reveal a genetic record of ancient farmers who were not the original hunter-gatherer peoples of Europe. Around 12,000 years ago, as the Northern Hemisphere warmed at the end of the Ice Age, a wave of trader-colonizers from the coastal Levant migrated northward, bringing goats, sheep, and precious seeds of emmer and einkorn to new lands. As grain growing became central to the sustenance of Neolithic peoples, the farming society generated an outpouring of megalithic sacred circles for seasonal ceremonies, combined with domestic altars and ritual figurines that all reflect a constant aligning of human communities with the vibrant forces of nature and the seasonal cycles of agrarian life. A deep devotion to sacred practices was at the center of early farmers' belief system.

THE GREAT FLOOD OF BREAD WHEAT

The Caucasus Mountains are one of the oldest centers of human habitation anywhere in the world outside of Africa. The spectacular geography of fertile coastal fields and mountains generated an abundance of biodiversity unparalleled in any other region. In the foothills of Caucasus Georgia, advanced flax weaving and needlework radiocarbon-dated to 32,000 years ago has been found.[18] Over 9,000 years ago early farmers used razor-sharp black obsidian tools; raised goats, cattle, and pigs; and grew a vast diversity of food crops including grapes, creating the oldest traditions of winemaking in the world. Archaeologists have unearthed distinctive artwork showing considerable skill in weaving, dying, stoneworking and metallurgy.[19] To this day in Georgian rural villages, ancient polyphonic chants and rituals celebrating weather deities, round dances dedicated to the sun, curative chants to nature's powers, and age-old seasonal ceremonies reflect the earth-centered tradition of earlier times.

As the Ice Age glaciers melted, sea levels rose worldwide, and the land bridge near modern Istanbul burst open. Vast fertile lands under today's Black Sea became submerged. Mediterranean saltwater poured over at a rate more powerful than Niagara Falls, covering more than 100,000 square kilometers of fertile coast farmland with seawater, displacing Black Sea coastal farmers. Where did they go?

Ancient peoples of today's Georgian and Bulgarian coast fled to the higher grounds of Europe, bringing farming, horses, iron forging – and a revolutionary new type of hull-less grain.[20] This wave of Neolithic farmers arrived into Europe around 7,500 years ago, fleeing the Great Flood of melting Ice Age glaciers. The Caucasus farmers (and later ancient Greek seafaring traders) brought a prized hull-less wheat with free-threshing kernels that revolutionized Neolithic farming. Known today as bread wheat, this ancient Caucasus wheat quickly spread throughout Europe's small farms, replacing the hulled emmer and einkorn grains from the Levant.[21] The bread wheat (*T. aestivum*) brought by ancient Caucasus farmers to

Europe as they escaped the inundating Great Flood is the most widely grown species of wheat in the world today.[22]

How the First Farmers
Imprinted Ancient Grains

With scientific evidence suggesting that plants respond to human energies, we can approach the evolution of wheat in the context of the belief systems and farming practices of traditional cultures with greater insight. The most ancient religion of humankind is the goddess religion. Archaeologists have uncovered life-affirming, earth-centered goddess civilizations of Old Europe's earliest farming cultures; the period when ancient grains evolved and were selected from wild plants into food crops.

When was the beginning?[23] Archaeologists have found early human artifacts, cave paintings, and a goddess figurine dated long before recorded history from over 40,000 years ago in Old Europe to at least 232,000 years ago in the Levant.[24] Long before Mesopotamia or Egypt, from the seventh millennium BCE — at least 8,000 years ago — early farmers in Europe lived in organized villages and created sophisticated art reflecting a peaceful, egalitarian society embraced by a nurturing female cosmic being. A profusion of designs on ceramics and sculptures from the sixth millennium BCE depict rhythmically interconnecting spirals, zigzags, circles within circles, forms coiling and uncoiling, interweaving plants, especially wheat plants, and animal and human forms. The perception of a cyclic universe orchestrated by mystical forces was the core worldview of the old culture.

Interconnecting motifs were created by indigenous peoples on every continent, who shared a sacred relationship with the living world. These designs echo the beliefs of quantum physicists, who describe the universe as a web of relationships between the various parts of a unified whole. Physicist Fritjof Capra identifies dynamic patterns on the micro and macro levels that continually change into one another as a "continuous dance of energy."[25] Scientist David Bohm sees the universe as a hologram in which the entire cosmic web is enfolded within each of its parts.[26] In ancient times the sense of reality was different. There was a deep connection between humans, animals, plants, and inanimate objects. This was a period when we were not separate from nature. Unlike modern humans, people in the ancient world perceived the cosmos, the stars, the earth, and life itself as parts of a whole. People thought that the separateness observed in the sky and on earth was an illusion, and that wholeness was the ultimate truth. People believed their myths were as true an organization of the universe as we believe our scientific explanations to be.

Early farming societies show no evidence of warfare. Gimbutas found a ratio of 98 percent female goddess figurines to 2 percent male figures.[27] Although abundant figures of female forms were found, there are no weapons in any grave or village ruins. The rich legacy of paintings, figurines, and sacred sites represent the goddess at the center of a harmonious, hardworking life in complex settlements of up to 50,000 people. Fascinating research by Dr. Harald Haarmann, a world expert on written languages, confirms that these people developed the earliest writing 8,000 years ago, millennia before writing developed in Mesopotamia or ancient Greece.[28] The households, trading system, temples, and graves in Old Europe document cooperative, peaceful, sophisticated societies with a female, self-regenerating goddess, giver of life and wielder of death, the unity of all nature in water and fire, stone and earth. She was sacred. We wonder today: how did this reality inform the plant breeding, imprinting on the seeds themselves?[29]

SACRED HEARTH

The bread oven was a principal feature of ancient temples in Old Europe. Shrines contained figurines grinding flour, kneading dough, and baking bread. Throughout Old Europe, especially in the Cucuteni-Trypillian culture from 8,000 years ago in modern-day Balkan lands, female figurines were found next to the ovens, on the ovens, where the grains were prepared, and in the grain storage bins. The earliest evidence of these, dating to 6500 BCE, was found in northern Greece. "In Thessaly they had rituals before and during the baking of bread. Women used small figurines as they were making the bread. Bread was sacred."[30]

Pregnant goddesses were worshiped in the courtyard near bread ovens. Organic remains of plants or grains were found that were sacrificed. The prehistoric bread oven was a sacred space of the Grain Mother. This can be seen from the anthropomorphized miniature ovens found from Malta and the Danube Valley to the shores of the Black Sea. The Goddess's eyes are above the oven door, which forms her mouth. Many ovens are decorated with spirals, as are the goddess figurines,[31] with yin-and-yang spirals thousands of years prior to their emergence in China, or anthropomorphized "womb" ovens in the shape of a pregnant woman with an umbilical cord on the top. Ovens are engraved with energetic spiral lines. Dough prepared in the temples was sacred bread used in life-affirming worship. Breads marked with spirals of life may have been the first offerings to the earth fertility goddess.

Since primordial times the oven has represented the womb, death, and regeneration. Baked bread evokes both the fertile earth and the place in the earth where one's body is buried. The symbol of the oven-womb as a portal to death and rebirth is retold in Old European fairy tales about Baba Yaga, the fierce, wise woman-sorceress of the deep forest. Baba Yaga is the wild forest spirit that guides the seeker on his path, or ensnares and drags him into the oven to be consumed by death. The strong and clever escape the fire, while the foolish are captured. Baba Yaga flies in a mortar and pestle, the symbol of the womb and the phallus; her home spins like the cycles of the moon; her oven is the hearth of primordial fire, transmutation, and protection. Baba is also the name of the life force in the harvested grain; the last moment of the harvest consecrated in the holy sheaf dedicated for new planting in spring.[32]

The pregnant belly of the Mother Earth was identified with natural mounds of all kinds, from hills to bread ovens, even bread itself. During the harvest, at the end of the scything, the top of the hill was left uncut and cleaned of weeds. The owner tied a knot of stalks of grain, since the last sheaf was considered the mound's umbilical cord. In eastern Europe, the mower who cut the last sheaf was the "cutter of the umbilical cord." Sacred hills were venerated since the earliest days. The worship of the mother was celebrated on mountains, especially ones with large stones. The festival of Lughnasadh was celebrated on sacred hills where the people garlanded large stones with harvest bounty. Folktales have kept this memory alive in legends of how the young seeker enters the magic mountain and finds within the beautiful maiden who holds the secrets of nature, plants, and the water.

GRAIN MOTHER

Throughout Europe to Asia, peasants believe that the grain field is animated by a soul. The life processes of reproduction, growth, death, and decay are the same principles that are seen in human beings. In the plant, as in the human, there is a vital element, a soul of a plant that is like the vital force of the soul. This belief in the plant-soul is at the heart of age-old cereal traditions. The spirit of the grain is the Great Grain Mother.

The Grain Mother spirit personifies the life forces in nature that make the crop grow. These traditions hark back to the dawn of agriculture before organized priests or priestesses but were rites performed by all the village farmers and their families. Villages each developed unique local traditions to honor and celebrate the Grain Mother. There were no temples or hierarchy of priests but rather sacred places in nature. Spirits were celebrated, not gods. The fertility of the earth or timely rains were evoked by ceremonies, which were believed to influence nature directly through a physical sympathy or resemblance between the rite and the effect, which it is the intention of the rite to awaken, not by propitiating the favor of distant divine beings but through sacrifice and prayer, celebration, praise, and attunement.

The farmers behaved toward the grain in bloom as they would toward a pregnant woman. They abstained from making startling noises in the field, lest they frighten the soul of the grain, causing it to miscarry and bear no grain. For the same reason they would not talk of anything negative, such as death or demons, in the fields. They imagined what her feelings must be at harvest, when people must harvest the grain with a knife, and took every precaution to reap in as painless and swift a manner as possible to not frighten the grain spirit till the very last moment. Although she lived in the fields during the stages of growth, she was believed to be caught in the last sheaves left standing in the field that were then joyfully carried home dressed in white and honored as a sacred being.

In Slavic eastern Europe, ancient harvest folk rituals were practiced that mark the opening and closing of the harvest period. Ceremonies interacted with natural processes and phenomena with the intention of the spiritualization and awakening of nature forces to influence the outcome of the harvest.

The beginning of the harvest was opened with a special ceremony. In the morning, all the reapers went into the fields together. The village elder took off his hat, turned to the sun, and uttered a special incantation requesting the fields to surrender their harvest and to give the reapers sufficient strength to gather it. In recent times, the incantation was replaced with a prayer. An honored woman cut the first sheaf of grain. Women reapers rolled on the field to absorb strength and fertility from the earth. The honored woman presented the elders with the first sheaf, after which everyone was offered liquor and festive foods. In the evening, the first sheaf was brought to the elder's house or the church altar.

At the end of harvesting, a stand of unreaped wheat or rye left at the edge of the field was tied into a sheaf known alternatively as Elijah's beard, the Savior's beard, Grandfather's beard, or simply the beard, and was left as an offering to the gods of the fields. In some regions the "beard" was thought to contain the benevolent spirits of ancestors who would protect the fields. Thereafter the reapers walked in a procession from the fields to the elder or master's house, carrying the last sheaf with wishes of happiness and abundance. The elder or master carried the ceremonial gift into the house and placed it in the altar corner. The harvest feast ensued, whereupon everyone enjoyed ritually decorated wedding-type breads, drinks, and food with traditional harvest songs followed by dancing, singing, games, or sport competitions.

In the dark of winter, when the days begin to lengthen, the festive winter solstice or Christmas Eve family meal was opened by the father taking a sheaf of wheat, dipping it into living water, and sprinkling drops of the water throughout the home and on the oven to protect from troubles and fire.

GRAIN MOTHER SEED SAVING

In Old Europe, the Grain Mother was fashioned from the finest and largest seedheads, often gathered by the elder women in the village. In *The Golden Bough*, James Frazer reports that the sheaf was often made into the shape of a woman by a *baba,* the oldest married woman in a village. The finest ears were plucked out of it and made into a wreath twined with flowers. This harvest crown was carried on the head of a young girl of the village to the farmer, or in later periods was placed on the church altar. In other villages the Grain Mother was carried at the top of a pole behind the girl who wore the crown. The Grain Mother was then placed on a high table in the center of the harvest supper, surrounded by festive dancing. Afterward the Grain Mother was hung in the barn till threshing was over, then kept in the home in a place of honor till spring. In spring, when the grain would wave in the wind, the peasants would say, "Here comes the Grain Mother," or "The Grain Mother is running through the field."[33]

Sacred Mother Sheaves of the best ears were mixed with the seed wheat for planting. The fertilizing power of the Grain Mother was awakened by scattering the seeds from her body over the newly planted field. At planting time she was brought to the fields to fertilize and bless the crops; however, if she was angry with a farmer, she might wither his grain. Often a young maiden rubbed out and scattered the sacred grains at planting time. After the field consecrations, the seeds were plowed into the field to fertilize the new planting for a good harvest and drenched with water as a rain-awakening evocation.

These seed-saving traditions continue to this day in rural villages. When I lived in the Middle East, villagers proudly explained to me that each year the grandmothers of the village went out to the fields to collect the finest sheaves of wheat to save for seed.

Female forces associated with the earth and the mystery of creation are expressed in the baking of bread. From ancient days, the production of bread has been associated with the forces of life. Leaven, the natural wild yeast, is called mother in English and madre in Spanish and Italian. In France a young woman pregnant before marriage was said to have "borrowed" a loaf. "A bun in the oven" also refers to pregnancy. The French word *four* (or an earlier word, *forn*) for oven is from Latin *fornicatio*, derived from *fornix*, which literally meant a vault but figuratively meant a prostitute. The French word *miche*, a round bread loaf, means breast or buttocks. In English, buns refer to the buttocks as well as round rolls. In Germany *Brotleib* refers to the female body. In Sicily, a festive Easter bread represents woman's fertility, *pupu cu l'ova* — bread made in the form of a woman pregnant with colored eggs. The eggs may also be made of fine chocolate.

FARMERS' MEGALITHS

As the Northern Hemisphere warmed at the end of the Ice Age, trader-explorers and pioneer-colonizers from the Levant migrated northward, bringing goats, sheep, and grain to new lands. The rise of farming stimulated the building of megalithic stone calendar structures that aligned the farmers with the vast cycles of the solstice and equinox. Neolithic early farmers generated a plethora of megalithic sacred circles for seasonal ceremonies to attune human communities with the vibrant forces of nature and the cycles of planting, tending, and harvest. A deep reverence for nature was at the center of their farming practices. Over a thousand megalithic stone calendars used for ceremonial farming and nature attunement and for predicting solar and lunar eclipses were designed using complex mathematics.[34]

Venerable Stones

Legends abound of great stones containing natural holes with healing properties. Rainwater falling into the holes acquired magic properties. Peasant women stopped by the stones to cure their aches and pains by washing themselves with the water. Large stones were often carved with symbols of suns and snakes. In Lithuania there were stone monuments usually about 6 feet high, smoothly cut, and surrounded by a ditch dedicated to goddesses who spun the fates of men. Great stones with flat surfaces were called goddesses. These great stones were covered with straw and venerated as protectors of crops and animals.

Sacred Waters

Rivers and lakes venerated in antiquity are often considered sacred in folk traditions to this day. The purity of water was cherished. No one dared pollute the life-giving water that had healing and fertilizing properties. If one sprinkled this vital water on plants, the flowers and trees would blossom. Sacred water was spread on fields to ensure good crops. Animals were anointed with it to keep them healthy. Washing with clear spring water was believed to heal eye and skin diseases. At the beginning of summer, during the sun festival (present-day Saint John's Eve or Midsummer's Eve), people immersed themselves in holy waters so that they would be healthy and beautiful, which encouraged young people to marry.

Purifying Fire

Peasants are great venerators of fire, the purifying element. Fire was sacred and eternal. Tribes had sanctuaries on high hills and riverbanks where fire was kept, guarded by priests. In each house a sacred hearth burned in which fire was never extinguished. Only once a year, on the eve of the midsummer festival, was the fire symbolically extinguished and then rekindled. Fire was a goddess who required offerings. She was fed, carefully guarded, and covered over at night by the mother of the family. Baltic peasants call this flame "mother of fire."

Lughnasadh:
The Celtic Grain Harvest Festival

Let the people use this day to make offerings to the Mother Goddess river of their region and to bathe themselves and their farm animals in "living wild water" — a stream or a river, a lake, a pond, or the sea. In every case shall offerings and prayers of thanksgiving be made.

Then shall the people make offerings to standing stones, dressing them with wreaths of new grain and flowers, placing first fruits at their feet. Let them cut a sheaf of new grain and set it afire in the ancient manner, burning off the husks with fire. Then let them grind the grain in a quern or a mill and bake a bannock from it saying:

On the feast day of Lugh and of Danu, I cut a handful of the new grain, I purified it by fire, and rubbed it sharply from the husk with my own hand. I ground it in a quern, I baked it on a fan of sheepskin, I toasted it to a fire of rowan, I shared it round my people. I went sunwise round my dwelling. In the names of Lugh and Danu, who have preserved me, who are preserving me, and who will preserve me, in peace, in flocks, in strength of heart, in labor, in love, in wisdom and mercy, until the day of my death.[35]

— Máire MacNeill, The Festival of Lughnasa

The beginning of the harvest was one of the four great festivals of the year. There was a ceremonial reaping of the grain by the head of the household, waving it thrice above his head, chanting the blessing that invokes protection from ills that threaten the crop. In ancient times the head of the community or the king performed the first cutting. The grain was threshed, winnowed, and ground. Ritual bannock, or griddle cakes, were made from the grain of the first harvest for each member of the family and enjoyed at festive communal feasts. Sharing food protected the people against starvation in the coming year. An offering was made by bringing the first sheaves to a high place and burying them. In ancient days a bull was sacrificed. Garlands of wildflowers decorated a maiden's hair, wreaths were placed on special places, flowers strewn on the summit and on the great stones. A girl was seated on a chair, garlanded with flowers. Berries were picked. Great dancing festivities ensued on the high places. The young men competed in tests of strength and agility. The people assembled by lakes and riversides, immersing themselves, their horses, and their cows in the wild, pure, holy water.[36]

The matriarchal, earth-based society went underground when fierce Indo-European warriors from the Eurasian steppes rode into Old European farming villages, pillaging and intermixing, causing the collapse of the Old European matriarchal culture and replacing it with the patriarchal culture of the nomadic warrior.[37]

Thus Planted Zarathustra

The Zoroastrian holy book Avesta recounts the legend of an Evil Killer Winter of 20,000 years ago. The Creator warned the good King Jam that a Killer Winter was approaching. All living creatures would be destroyed unless his people traveled far south. The Great Migration of the ancient Zoroastrians from their northern homeland is recounted in *The Saga of the Zoroastrian Race* by Porus Havewala.[38] Another Zoroastrian legend retells the profound farming skills of Zarathustra:

Long ago, in the distant past, there was a great king in Iran whose name was Vishtap, with great wealth and victories to his name. One day, Vishtap was traveling through his kingdom and passed an orchard of exceptional beauty, although the land around the orchard was bare and desolate. Vishtap realized that the farmer tending the orchard must have worked with great foresight, wisdom, dedication, and diligence, since he had never seen such beauty.

Vishtap asked who was the gardener. He was told that this was the garden of Zarathustra. He invited Zarathustra to his palace to ask him deep questions of life, however, instead of sharing his secrets, Zarathustra reached into his satchel and gave the king grains of wheat, saying that the grains were his teacher. Zarathustra then excused himself, explaining his work in the orchard could not be neglected. The king was surprised by Zarathustra's response, but wondered, perhaps the grains had magical powers? The king placed the grain securely in a gold box. Every day he opened the gold box and looked at the grain to find the answers to his questions. Every day he was none the wiser.

After months of frustration, the king visited the sage again. The king showed Zarathustra the grain in the gold box, and asked him what lesson the grain was to teach, whereupon Zarathustra asked the king what would have happened if instead of placing the grain in a gold box, the king had planted the grain so that it could receive the forces of life, food, water, and light. Together they reflected on the deep lessons they could learn from that simple act. To grow and transform, the grain was removed from the gold box and grounded in the earth. Thus the king understood that he too must step out of his comfortable surroundings, as the forces of nature would flow toward the grain to nurture its growth, so would he be nurtured by life with knowledge and understanding.[39]

The Zoroastrian holy book Vendidad, which is part of the Avesta, teaches, "who causes wheat to be sown, causes righteousness to be practiced."

ANCIENT PERSIAN WHEAT (*TRITICUM CARTHLICUM*)

Ancient Zoroastrian traditions are almost forgotten today, but Zarathustra taught practices to achieve a deep harmony with nature, drawing on secrets of the plant and animal kingdoms as the people farmed. They communicated with nature in seasonal festivals called *gahambars* as a way to express gratitude to the plants and animals. Gahambars were performed in open high places, and they sanctified relationships with the nature beings: Spenta Armaiti, who is responsible for the fertility of the earth; Khordad, who looks after the waters; and Amordad, who looks after vegetation and crops.

Communication with plants and animals is natural. If we think we need supernatural powers to be able to communicate, it is only because our modern way of life has become dependent on mechanisms, causing natural abilities to be forgotten. A simple example is the use of calculators in school. Until recently, calculators were not allowed in schools, yet today students cannot imagine doing math without a calculator. Scientific experiments conducted with plants document that singing praise to plants gives better fruits and playing music to cows doubles their milk. These practices are popular because they increase wealth, however in ancient times, farmers praised nature because it was the good way. The Zoroastrian "Alaat" are sacred vibratory prayers to attune with nature and cleanse energetic blockages, awakening our divine connection with natural creation.[40]

Biblical Grain Traditions

The biblical ideal of the human being is the small-scale farmer. Decentralized. Self-sufficient. Local. The ancient Israeli religion was developed by and for small-scale organic farmers. When the Jewish nation entered the land of Israel, the centrality of food, bread, and the produce of the land in the life of the people evolved both in the context of and in contrast to Canaanite practices and the top-heavy power system of ancient Egypt. From the moment that the Jews crossed the Jordan River into the land of Israel, the women were instructed by Moses to offer up a portion of their daily bread dough. It was the women who saved the seeds, baked and sanctified the bread, and brewed the beer.

> Rabbi Ahai ben Josiah said, "He who buys grain in the market, to what may he be compared? To a child who is cut off from his mother, and although it is taken to homes of wet nurses, it is not satisfied. And he who buys bread in the market, to what is he compared? To a man who digs his own grave — a wretched, precarious existence. But he who eats of his own produce is like a child reared at his mother's breast.
> — Fathers According to Rabbi Nathan (*Avot d'Rabbi Nathan*) 30:6[41]

FOOD JUSTICE

During the 1,470-year period from when Joshua and the Twelve Tribes conquered Canaan in 1400 BCE, to the conquest of the nation of Israel by

In the Fields of Tekoa

In the hills of Tekoa, the farmers stood amid the wheat fields. Some swung their scythes among the stalks and sang: "They that sow in tears shall reap in joy"; others bound sheaves, singing: "He that goes forth and weeps, bearing precious seed, shall come again rejoicing, bringing his sheaves with him." Still others gathered the grain into the threshing floor and sang, "My standing wheat knelt and bowed to my sheaf. Lord! Bless my strength to gather my grain." The field was filled with work and song.

Eliav, his wife, and two sons came to their field to begin the harvest. They came to the standing wheat from the four corners of his field. Swiftly glittered the scythes in the hands of the nimble workers. The stalks fell hither and fro as the heads of the reapers were seen above the golden sea. A neighbor called out to Eliav:

"From your standing wheat to your threshing floor, my neighbor." "From your vineyard to your wine cellar," replied Eliav. "The blessing of the Lord on you," called out Eliav's wife to her neighbor. "Have you washed the wool of your flocks?" "The Lord be with you, my neighbor," she replied. "I have washed it and carded it." "I have already woven thread." "The Lord sends us blessings according to our abilities!"

"Zizz! Zizz" whispered the scythes to each other. It was as if they had scolded and said: "Hush, women. Don't gossip so much. Time to work." The women cut short their conversation and swung their scythes in the grain. "Zizz! Zizz!" whispered the scythes to each other. At that moment there passed a wanderer from the land of Moab. He was tired and hungry. Eliav saw him and called out: "The Lord be with you, wanderer! Turn hither, and pluck for yourself ears from the wheat, and bless the Lord that he has sent us his blessing." And the wanderer came, and plucked wheat, ate, and was sated. "Wait, wanderer," Eliav said to him. "Behold, I am cutting my wheat; take from the corner." But the wanderer did not understand the significance of the word *corner*. Eliav said to him: "This is the custom of the Hebrew farmers: when we reap our wheat, we leave stalks in the corner of the field for the poor and the wandering. It is called the corner." The wanderer plucked wheat, rubbed it out, placed the seed in his pouch, and went on.

The wanderer passed the field of Eliav's neighbor, and a voice called out behind him: "O, blessed of the Lord! Why do you shame me? For I have done you no wrong." And the wanderer was taken aback and said: "O, my lord, when did I shame you? I am a stranger. Only now have I come from the land of Moab, and I have never seen your face until today." "And do you not shame me," replied the farmer, "when you pass my field while I am binding the sheaves and do not gather the gleanings?" "What are the gleanings?" asked the wanderer. "It is the way of the Hebrews," answered the farmer. "The reaper grasps a handful of stalks, and the scythe cuts them below. The stalks slipped from the hand and escaped the scythe; they are not for the reaper. The Lord has saved them for the poor and the wandering."

The wanderer placed the seed in his pouch and went on. A voice called to him: "Stranger! Will you do thus to me? Behold, I am making a threshing heap, and will you not turn to me and collect the forgotten?" "O, my lord, I know what are the gleanings and the corner, but I do not know what the forgotten is."

And the farmer said: "This is the way of the Hebrew farmer. When a man gathers his sheaves to the threshing heap, and has forgotten sheaves behind him, it is a sign that the Lord has given them to the wandering and the poor. Now you go and pass through my field, and you will seek and find sheaves. Take what the Lord has saved for your sake."

The wanderer gathered the forgotten sheaves, beat them out, rubbed them, placed the seed in his pouch, and went on his way.[42]

Rome in 70 CE, Israel developed a sophisticated community food system to ensure food justice on a scale unheard of in the ancient world. What began as a revolutionary escape from oppression evolved into a society organized by peasant farmers.

The Talmud documents a society that elevated growing food into a sacred practice. Just as today the organic farmer understands we need to nourish the soil and the earthworms in order for the earth to be fertile, the ancient Israelis not only enriched the soil with animal waste, compost, ash, dried blood, fallowing, and crop rotations but believed that it was necessary to feed the people in order for the earth to be fruitful. Ancient Israelites believed that soil fertility was based on food justice.

> We understand that the entire earth globe is one individual being which is endowed with life, motion and a soul. Not as some persons maintain, inanimate matter like the elements of earth, fire, air or water, but an animate, organized, intelligent, moral being capable of comprehension and response.[43]

The ancient Israelis believed that the earth — a living, conscious being — would be fruitful as long as the farmers nourished the community, the poor, widows, and orphans; that by feeding the people — all the people — the earth would provide its bounty. The food system of the ancient Israelis involved seasonal temple consecrations of planting, growth, and harvest with community food gifts from the field, vineyard, and orchard before they could be enjoyed by the farmers or sold. A sabbatical year, or *shmita*, was celebrated every seventh year wherein all produce was communal, wild-gathered foods were the staple, and all debts of the farmers were absolved. Any land sold by a farmer to pay off a debt reverted back to the farmer. Shmita was not an agricultural fallow, since biannual fallows were already practiced. It was rather a seven-year cycle of spiritual renewal, a restoration of justice for the living earth and her creatures.

Millennia of indigenous farmer knowledge on how to cultivate healthy, disease-free wheat is embedded in Jewish food and farming traditions, passed down to this day in the Talmud. From the spring cleaning of grain storage areas that rid the home and barn of pathogens and insect vectors, to the *shmurah*[44] watching of the grain field to prevent contamination from moisture at critical times, these ancient farmers sure knew something about wheat cultivation. The extraordinary practices of community food systems documented in the first volume of the Mishnah, titled *Seder Zeraim*, or the "Way of the Farmer," were written down in the second to the fifth century CE, with ensuing centuries of rabbinic commentary in the Talmud.

THE HEBREW GODDESS

For the first 3,000 years of Judaism, the people worshiped the Goddesses and a tribal male deity. The religious practices of ancient agrarian peoples evolved to enhance the soil fertility and invoke the rainfall on which life depended. From the time when the Jews entered the Holy Land in 1350 BCE until the Babylonian conquest in 587 BCE (over seven centuries) the Hebraic culture embraced a cosmology in which goddess worship was normative. Judging from the frequent occurrence of the goddess figurines not matched anywhere by any images of a male god, the worship of the Goddess was extremely popular in all levels of Hebrew society.[45] During the First Temple period from 957 BCE to 586 BCE, goddess worship within the Temple itself was sanctioned, legitimate worship. The Asherah living tree stood within the temple for most of the First Temple period. The Hebrew goddess was worshiped not as a foreign cult but as an embedded expression of Hebrew practice, symbolized by the tree of life. The goddess was worshiped in almost every home in ancient Israel.[46] The life processes of wheat, from tending the soil to food distribution and baking bread, were her expression.

After the destruction of the First Temple by Babylonians, and the ninety-year exile of the Israeli elite in Babylon, the rebuilding of Jerusalem Temple was financed by King Cyrus the Great of Babylon. Zoroastrian-influenced Nechemia and Ezra the Scribe compiled ancient Hebraic legends and fragmentary texts into a unified bible known today as the Torah. They introduced Zoroastrian-inspired monotheism to the people of Israel as a means to unify the empire of Cyrus the Great, evolving a rich synthesis of biblical traditions. During the Second Temple period, the goddess tradition was forced underground by political pressure from the male-dominated priesthood.

Although the Goddess was no longer recognized as the primary being, the feminine spirit became known as the Shekinah, the sacred feminine expression of holiness. Ancient Hebrews continued to honor sexuality as a sacred, life-nourishing gift. In the Holy of Holies, the inner sanctum of the temple in Jerusalem, an image of a male and a female angel in an intimate, marital, sacred embrace adorned the Ark of the Covenant.

Said Resh Lakish, "When the Gentiles entered the Sanctuary, they saw the cherubim joined together in a sacred embrace. They took the figures to the marketplace and said, 'Should these Israelites who are so close to G-d, whose blessing is a blessing and whose curse is a curse, be involved in such erotic matters?' Immediately, the Romans debased the Israelites, as it is said, (Lamentations 1:8) All who once respected her (Israel), debased her, for they saw her inner nakedness exposed."[47]

The Hebraic symbol of interpenetrating triangles, known as the Star of David or Seal of Solomon, is an ancient fertility symbol, known to the Hindus as the Great Yantra. It represents the "Great Rite" of sexual intercourse, merging male and female energies. *Yihud,* the Hebrew word for the Great Rite, literally meaning "intimacy or unity," refers to the process of bringing together the complementary male and female principles into a dynamic harmony. It can also refer to a psychological balancing of one's inner male and female aspects.

The Shekinah Divine Presence rests on the marital bed. There is nothing more holy and pure than when a husband and wife are intimate for the sake of Heaven. After destruction of the Temple in Jerusalem, the bedroom in a loving home is the place of the Holy of Holies.[48]

Adam and Eve were created equally intertwined within one another, *panim le' panim,* face-to-face, as symbolized by the images of the Cherubim on the Ark.[49]

SACRED EMBRACE

The tender embrace between two people transforms as you turn it in your hand; a swollen penis from the back to full breasts from above and a vagina from below, evoking the act of making love in its primal, intimate fullness. Even more profound is not just that there are two human figures embracing but that they are equal beings. Male and female merge into the intimate embrace itself, suggesting not only reproductive vigor but tender face-to-face intimacy. At the moment of the emergence of agriculture is the power of sacred sex, its creative force imbuing the life processes of farming.

The Natufian fertility figurine, shown on page 8 in the color insert, was fashioned by the first wheat farmers 12,000 years ago, discovered in

1933 in Wadi Haritoun, Tekoa, near Bethlehem, in the very same village where I lived in Israel. The figure was among household objects found in a cave dwelling. It was part of the daily life of the first wheat farmers. The Natufian culture flourished in the eastern Mediterranean Levant from pre–Ice Age hunter-gatherers to around 10,000 years ago, when, as the climate warmed, farming emerged. Sickle blades appeared for the first time. The characteristic sickle gloss on the blades indicates that they were used to cut the silica-rich stems of cereals, which is evidence for early agriculture.

Klaas Martens looked thoughtfully at this figurine when I showed him the photo. He looked deeply at me, his comment explaining everything: "This is the wheat itself" — the intimate self-pollinating embrace of the first farmers within the soul of the first wheats.

I know Wadi Haritoun well. I lived there, spending fifteen years gathering seeds of wild arugula, lettuce, chickpea, lentil, artichoke, wild mallow, barley, emmer, and einkorn in the wild meadows of Tekoa. I replanted them in the fertile terraces that I built with my own hands, after years of pulling out the rocks, digging deep into the compact soil, enriching the heavy clay with manure and straw, gathering, planting, then harvesting the seeds in their fullness. This garden was embraced by olive, pomegranate, almond, and fig trees garlanding the wild landscape. Fragrant wild thyme, sage, and hyssop herbs were tenaciously rooted in rocky crevices. After my first year of collecting and planting, the wild seeds easily naturalized in the enriched, deeply dug soil, becoming a perennial Garden of Eden, ever-bearing, ever-abundant. My own little "Neolithic revolution" of agriculture took root over brief seasons of gathering and replanting. A few hundred feet from my homestead was an Arab village, still herding goats, still growing the very same ancient durum and barley, and baking in the *taboun* oven of biblical days.

The primal connection of the fertility forces of the Earth Mother and the baking of bread became vital for me during the years I spent in the homes of traditional farming families. I worked in their fields, played with their children, attended their weddings, and was accepted into the women's inner realm of the kitchen, where few outside males or Western researchers enter. The bread-baking traditions in the villages surrounding Jerusalem have been passed down from generation to generation. I always asked the women to teach me how to bake their delicious bread. In home after home,

I experienced the deeply erotic bread-baking process as the women knelt on the ground and kneaded the dough with their whole bodies in undulating rhythm. The bread was often baked in a clay taboun oven that echoes the shape of the breast, or a womb of warmth, as the bread gestates. The regenerative life forces of bread baking are vital, palpable in these traditional farming homes.

Some of the families I worked with are descended from Jewish farmers who stayed on the land after the Roman conquest 2,000 years ago. These families were forced to convert to Islam in the ninth century. The baking methods from such a family, described below, echo the baking of ancient Israel.

The Arabic bread blessing phrase, *Bismillah ir-Rahman ir-Rahim* means, "Blessed be the place of utmost tenderness that gives nourishing protection." The word *Rahman ir-Rahim* means "womb" in Hebrew and Arabic.[50]

Wheat and War

The land of Israel is a narrow, long corridor bordered on one side by the cool breezes of the Mediterranean and on the other by scorching desert. This corridor, connecting the ancient civilizations of Egypt and Mesopotamia, was for thousands of years a battleground for warring empires and tribes, each bearing wheat varieties from all directions into the land.

Wheat was transported throughout the region by commerce internally and with neighboring nations, resulting in intensive intermixing of regional wheat gene pools. After the Roman conquest, by the third century the Byzantine-based Roman Empire controlled the land of Israel until desert Muslim tribes invaded in the mid-seventh century. After the Crusader invasions of 1100–1250, the Mamluk regime destroyed all seaports to prevent future invasions, indirectly preventing any wheat movement. From that period ten centuries ago, the wheat cultivated in the land of Israel evolved in highly localized microclimates. The land was sparsely inhabited by small hostile clans, each living in a territory of defined natural boundaries. Each clan was at constant war with the others. There was no known interaction between these hostile groups other than warfare. Each depended on the products of its own district that were cultivated for centuries on the same soils with no outside introductions. This gave rise to numerous landrace populations with a foundation of rich genetic variability, specific to localized microclimates, with early origins spanning northern Africa to ancient Mesopotamia.

After the Crusades, the region suffered great impoverishment. Trade ceased. The genetically diverse landraces had centuries to stabilize traits into localized populations, selected by farmers for preferred characteristics such as shape, color, or flavor. The landrace durum wheats that evolved in the southern Fertile Crescent encompass many unique local varieties that can be characterized: Nursit types with slender, translucent seeds; jaljuli types with large seeds and long seedheads; and Hourani types with short, fat seedheads from the Hourani plains.

Hestia's Hearth

Early Greek temples were sanctified hearth houses.[51] Each village, city, and state maintained a community hearth with a sacred fire. Even the Temple of Apollo at Delphi had its inner Hestia hearth where the sacred fire was continuously lit, tended by the king or his family. This *Prytaneum*, or community hearth, was the religious and political center of the community and government. *Psadista*, the ritual bread offering to the gods, was made of fine flour, oil, and wine.[52]

Ancient Greek maritime trade contributed greatly to the migration of bread wheat (*Triticum aestivum*) from its ancestral homeland in the Black Sea region near the Caucasus Mountains to Athens. According to Demosthenes, Athens imported 400,000 *medimnoi* (approximately 4,800,000 liters) of grain per year in the late fourth century from the Crimean kingdom on the Bosporus alone.

SACRAMENTAL BREAD

Warrior Romans conquered and destroyed, from ancient Israel to the Celtic tribal lands throughout Europe, brutally murdering many, including druid priests. Those who stood in the path of this Grim Reaper were ruthlessly cut down. Later, as Rome collapsed, it morphed into the male-dominated religious power system based on the Roman Empire. Grain folk traditions went underground or were incorporated within church concepts. Christianity absorbed the ancient reverence of the grain cycle into its liturgy, identifying the Son of God as the heavenly bread and a wheat cracker as the divine body. The Christian metaphor of the Mother echoes the divine fecundity of the Grain Goddess. The planted Grain-Son descends into the soil — that is, the underworld — and rises from the dead to be harvested and sacrificed each seasonal cycle. The resurrected Son embodies the grain as his reaping-death gave life to humankind. The Grain Mother embodies the miracle of transformation from grain into bread.[53]

In the breathtaking beauty of Tuscany at harvest time, at a heritage wheat conference, a Renaissance Madonna and Child shrine greeted me at the farm gate. The local parish priest explained passionately to the farmers and bakers. "Let there be no mistake. This glistening field of wheat is not just a symbol of the Son. The wheat field is the Christ!"

Al-Andalus, Jewel of the World

Husbandry is the foundation of civilization — all sustenance derives from it, as well as the principal benefits and blessings that civilization brings.

— IBN ABDUN

The Arabic word *filāḥa* means "farming" or "husbandry," and *fellāḥ* means "husbandman," "tiller of the soil," or "farming-peasant." *Filāḥa* also means "to thrive," "prosper," "well-being," or "happiness." The word is sung out from the minarets of every mosque during the call to prayer: *hayya 'ala 'l-falāḥ*, which means, "come to well-being," or "come to holiness." Husbandry, well-being, and worship are inextricably united.

Rome's collapse in the fifth century left a power vacuum filled with rival barbarian tribes vying for power, while the Church created a doctrine to assert control over the soul in which Greek science was blasphemous. In 641 Islam conquered Alexandria, Egypt, the seat of learning with the largest libraries in the known world. Arab scholars inherited the wealth of Greek knowledge. Mosques were built adjacent to schools, introducing literacy to all who partook.

Agriculture blossomed with the introduction of new cultivars, intercropping rotation systems, the abundant use of manure, and advanced irrigation systems. For the first time, individuals had a right to own, buy, sell, mortgage, and inherit land, in contrast to Roman oppression, under which vast monocropped estates were worked by slaves to produce Roman grain tributes. Under the Moors, the old Roman estates were divided into small farms, market gardens, and orchards. "The legal and actual condition of the over-whelming majority of those who worked on the land was one of freedom"[54]

Although home to diverse, creative peoples, Islamic Spain was not a model of multicultural harmony. Andalusia was beset by religious, polit-ical, and racial conflicts controlled in the best of times by its vast army. Its achievements are inseparable from its turmoil. From the 1100s, north-ern Visigoth-German Christians conquered Spain, taking Toledo's rich libraries. From the South, puritanical North African Muslims overran al-Andalus, destroying the liberal diversity of the "Golden Age," murdering the Muslim leaders and outlawing Judaism and Christianity. In response, northern Christian kingdoms fought, burned libraries, and conquered. Some 700 years after Abdul Rachman, Isabella and Ferdinand set up their new palace in the renowned Alhambra mosque in southern Granada. Tragically, most of the scientific manuscripts that remained were burned by the conquerers. In the dust of the refugees fleeing, the light of al-An-dalus was almost lost. Muslims and Jews who fled not only carried ancient traditions but nostalgia for the breadth of knowledge that flourished in al-Andalus. Among the precious few that remained is *Kitab al-Filaha* (Book of Agriculture), written by Ibn al-Awwam, an Arab farmer-scholar who lived in Seville, Spain, in the 1100s, at the same time as Moses Maimonides. The following excerpt is adapted from *Kitab al-Filaha*:

In the first aspect the farmer should care for the good disposition of the land, cropping patterns and application of fertilizer as the soil requires. Farmers should always be cheerful and in good humor when working with the crops. One of the most wondrous aspects of farming is in the ordering of time and the seasons so that each activity is done at its proper time. When it is done at another, the result is never as favorable. The use of the land is the chief factor in maintaining and modifying the nature of the soil. Natural vegetation improves the character of soil. Roots penetrate clay soil, making it more porous and able to absorb water. Humus enables the soil to retain moisture longer. In this way, naturally poor soil may gradually become rich with vegetation, eventually capable of producing crops. The best of all soils is the alluviums of river valleys because of the silty mud with which they are mixed. The moving water brings downward the rich sediment carried from the upland soil surface along with leaves and manure that decompose, enriching and refreshing the lower land soil. One should level the fields, placing stone walls on the face of the hills by which the soil may be entrapped to accumulate as it is swept down into the declivities. The most desirable soil texture is a balanced loam that absorbs and drains all moisture whether it came from the sky as rain or to the land by irrigation. It is advised to systematically mix the respective clay soil with sand, and sandy soil with clay to improve the texture, and to reduce runoff and erosion.

Wheat or barley greatly fatigue the soil when they are grown repeatedly without interruption. If then we do not want our soils to be exhausted, we must alternate crops to preserve the soil's productive power by sowing legumes that include lupine, beans, peas, lentils, fenugreek, vetch, clover or alfalfa, which can serve in the place of manure. Manure and compost are best used in irrigated gardens.

The Nabateans explain that good-quality grain is swollen, heavy, smooth, consisting of a shade pulling between yellow and red, the dominant shade. Various other shades were well reported, and the brown color was not as appreciated. The harvest is ready when the grains take on a light yellow color fading to white. Wheat is not to be fully mature, but still must have some sappy moisture so that it is like an "almost-hard dough." Baking methods in ancient times were more advanced. The Nabateans give detailed descriptions for choice flour, preparation of the sourdough ferment or leaven, and kneading of the dough.[55]

Muslim farmers introduced new plants, fruit, and vegetables in a three-year irrigated rotation system with legumes, transforming the European landscape where farming had been dominated by the ancient hulled emmer, einkorn, and spelt, or soft wheats from the moist Caucasus region in a two-year crop or fallow system. New crops, including sugarcane, rice, citrus fruit, figs, dates, apricots, cotton, artichokes, eggplants, lettuce, chickpeas, lentils, durum, and saffron, required changes in farming methods. These cultivars from hot climates needed the heat of summer, traditionally a "dead" season in a Mediterranean agriculture that had previously been restricted to crops grown in the rainy winter months. The new crops had to be irrigated, but the new summer growing season led to widespread introduction of crop rotation, with one summer and one winter crop. Previously in Roman and Byzantine agriculture the two crops in rotation had eventually depleted the soil of its natural fertility. Soil exhaustion was a motivating factor in Greek and Roman expansion. Moorish farming practices renewed the soil with organic manures, mulches, and minerals, and integrated grazing livestock and field crops.

The Moors introduced hard durum wheat into Europe. Durum is better able to tolerate heat and drought than the soft wheats it replaced. Millet, sorghum (which the Berbers brought from Sudan), and rye were also cultivated. More recipes have survived from al-Andalus than from any other medieval society.

Planting with the Moon

In 1513 in the city of Granada in al-Andalus, Gabriel Alonso de Herrera wrote,

Farmers should be mindful that sowing, as well as grafting and plowing, must be done if possible when the moon is waxing, preferably at the beginning of the cycle. The moon's two quarters of waxing are for activities related to growth, while the two quarters of waning are for activities related to consumption. The first quarter, characterized by heat and humidity, fosters the growth of seeds and plants far more than the second quarter, which tends to be hot and dry. During the two waning quarters it is cold and wet and affects seeds more than trees. If seeds are sown during the waning moon, they may not sprout. If they do grow they may not be as hearty and good as they would be otherwise. If good weather is a certainty harvesting should be undertaken when the moon is waning, preferably at the end of the day of a waning moon, as it is ideal for preserving the grain and makes it less susceptible to moisture-related diseases and infestations. The wheat should be harvested quickly as it becomes full and dry so that it is less susceptible to disease or pests.[56]

My own experience confirms that aboveground crops do best when planted in the first and second quarters, and belowground crops when planted in the third and fourth quarters. As it is said, "Aboveground in the Light of the Moon. Belowground in the Dark of the Moon."

The Gaunt Face of Hunger

From the fall of Rome in 476 CE to the sixteenth century, before New World crops such as potato, tomato, squash, and corn were introduced, the cuisine of Europe and the Mediterranean lands was based on cereals, then more cereals — particularly on wheat, barley, rye, and millet, supplemented with legumes, eggs, dairy, foraged and cultivated vegetables, and meat and fish for the elite few who could afford them. Beer, frumenty pottage (grain stew), and bread were the staple foods composing at least 50 percent of the diet of the common person.[57] The centrality of bread in sacramental rituals in the church highlighted its value.

A typical village included a castle or manor house for the lord, a church, and a mill surrounded by artisans and humble peasant housing and nearby cultivated fields bordered by a wilderness of forests and marshes. Each village was self-sufficient because it was cut off from the rest of the world. Roads were poorly maintained and threatened by robbers. Peasants were bound to the land, not free to leave. Lords, whose rights were granted by the king, demanded taxes, rents, and labor from the serfs. A medieval lord could not evict a tenant nor hire labor to replace him without legal cause. Land as a commodity to be bought and sold developed only in the nineteenth century, when monarchies were overthrown.

In the early period peasants used a two-field rotation with only half the fields in cultivation in any given year. Crop yields were low. In the Roman Empire, for every bushel of seed grain planted, about four bushels were harvested. In the Early Middle Ages this ratio dropped to one-and-a-half

Wheat Harvest Divided

One part cast forth for rent due out of hand, One other part for seed to sow thy land, Another part leave Parson for his tithe, Another part for harvest, sickle and scythe, One part for plowrite, cartwrite, and smith, One part to hold thy teems that draw with, One part for servant and workman's wages to lay, One part for fill bellie day by day, One part thy wife's needful things doth crave, Thyselfe and childe that last one would have.[58]

or two to one. This meant that half of a peasant's harvest was saved for seed grain to plant. The medieval farmer generally had a two-crop rotation: a spring field of barley, vetches, oats, peas, or beans, with a fall field of wheat or rye. Wheat and rye were used for bread, barley for beer. Hay and oats were fed to livestock.

Life under a Medieval Roof

From the Neolithic Age until the late medieval period most of the buildings in Europe had thatched roofs. Smoke-blackened from open hearth fires over generations, the thatch wheat straw has been well preserved from insects and decomposition, a window into ancient fields of wheat. John Letts, an archaeobotanist with a passion for wheat and thatch, has uncovered this little-known treasure of well-preserved heritage wheats that cover English thatched roofs to this day. Letts brought samples of medieval thatch to gene banks to search for viable seed equivalents and today grows and bakes a typical medieval bread. Letts reports:

I excavated over 200 ancient thatched roofs, exposing large quantities of well-preserved medieval wheat. These samples contain mixtures of "landrace" English rivet wheat (*T. turgidum*), bread wheat and rye which grew to 6 ft height, far taller than modern varieties, as well as barley and oats. Rivet wheat has not been grown commercially since the late 19th century. Rivet produces excellent bread and thatching straw as late medieval herbals and agricultural treatises confirm. Landraces evolved over centuries when crops were grown in heterogeneous conditions, year after year, from seed saved from the previous year's crop. Every plant in a landrace population is slightly different from its neighbor. Medieval cereals were diverse in straw height, ripening time and other agronomic traits. This diversity ensured that a portion of the crop yielded grain irrespective of drought, waterlogging, frost or disease.[59]

Famine and Revolution

Medieval Europeans ate about two pounds of bread each day, be it the dark bread of the peasant or the finely sifted white bread of nobility. The warm period of the first thousand years of the Common Era plunged into the bitter cold of the Little Ice Age from the 1100s to mid-1800s, bringing famine and disease. Scarcity clouded the horizon. As the Little Ice Age worsened, hunger turned to pestilence. The spread of the Black Plague was hastened by malnutrition. Villages in the Swiss Alps were destroyed by encroaching glaciers during the mid-seventeenth century. Rivers were frequently frozen deeply enough to support ice skating and winter festivals. The first River Thames Frost Fair was in 1607 and the last in 1814. Cool, wet summers led to infestations of fungal rot and mildew on the grain, and even to an illness called St. Anthony's Fire. Grain, if stored in cool, damp conditions, may develop a fungus known as ergot that causes a reaction similar to the effects of LSD. Whole villages suffered convulsions, hallucinations, gangrene, and even death. Famine led to bread riots all over Europe and was a catalyst for the French Revolution. Scarce wheat flour was supplemented with barley, rye, or bean flour.

Thomas Tusser, 1524–1580, an English farmer, poet, and grain merchant and author of *Five Points of Good Husbandrie*, introduces the qualities of grain below:[60]

White wheat, red rivet or white, for passeth all other for land
  that is light.
White pollard or red that is richly set, for land that is heavy tis
  the best you can get.
Wheat that is mixed with white and with red is the best you
  can get in the market man's head.
So Turkey or Purkey many do love, because it is floury as
  others above.
Grey wheat is the grossest, yet good for the clay, though worst
  for the market as farmers do say.
Much like unto rye his properties be found, coarse flour, much
  bran and a peeler of ground.[61]

Around 1000 CE, inspired by the thriving Moorish agriculture and a more effective moldboard plow that turned over the soil instead of just opening up furrows, the system expanded to a three-field system of barley, oats, or legumes planted in spring, wheat or rye in the second field in the fall, and the third field left fallow, with yearly rotations of the fields.

An eleventh-century granary filled with carbonized grains was discovered in the village of La Gravette, France. The greatest proportions of species were bread wheat, hard durum, and rivet wheat. Oats, barley, rye, and einkorn were found in smaller quantities. The stored grain may have been from taxes paid to the lord who owned the granary.

Class distinctions were reflected in the types of flour used in breads: manchet, which was finely sifted wheat bread often with butter, eggs, and milk for nobility and clergy that may have used beer barm (brewer's yeast) for leavening; and maslin (from the French *miscelin* meaning mixture) or ravelled bread, which was mixed, unsifted rye, wheat, or even barley sourdough bread with the bran left in. Maslin was the mainstay of the merchant and yeoman landowning farmers. Rye-wheat mixtures were often grown together as an insurance against crop failure. By feudal law the lord was obligated to bake the bread for his serfs in a public oven for which they were taxed.

Wheat in the New World

The first wheats grown in the New World were the drought-hardy durums of Middle Eastern origin, brought to Mexico in 1523 by Cortez and Spanish monks to use in sacramental bread for communion. Wherever the monks traveled, they planted wheat for the consecrated wheat wafer in the ritual of the Eucharist. Sacramental wheats have been found in isolated areas that adapted over centuries to harsh wild conditions where only the most robust wheats survived, developing hearty resilience over generations. When rain failed, wheats without deep roots died off. Researchers have recently collected these ancient wheats to preserve their biodiversity and their disease-resistant and drought-tolerant traits.

As farmers gain access to improved varieties or migrate to cities, sacramental wheats are disappearing. The farmers who still grow these old wheats report that they taste better and are more hardy than modern varieties. With the hope of conserving these rare and valuable varieties,

Ryaninjun

In the late 1600s, outbreaks of a virulent fungal disease known as black stem rust or "blast," smote the wheat fields of colonial America. Early colonists in New England combined easy-to-grow rye with Indian corn for the staple bread they called ryaninjun. Ryaninjun was wrapped with green oak or cabbage leaves to hold the low-gluten dough together and keep in the moisture. Small children gathered young oak leaves specifically for wrapping dough. For those who could afford it, rare wheat was mixed in the rye and corn to add some lightness to the heavy fare. Only the wealthiest people could afford full wheaten bread.[62]

researchers went on wheat-gathering expeditions in nineteen Mexican states, to collect the sacramental wheats before they disappeared. When a new leaf rust disease appeared in Mexico, scientists discovered that the sacramental wheats showed the greatest resistance to the blight.

As we face water shortages and rising temperatures due to climate change, scientists are looking to sacramental wheats as a source of drought tolerance. Field trials show some sacramental wheats have better early ground cover, quickly blanketing the soil and preventing moisture from evaporating. Others fill the wheat grain even under drought conditions, while some show better water uptake in deep soils due to well-established roots. If the wheats didn't have deep roots and it didn't rain, they died. The wheats that survived and were selected by traditional farmers over centuries were those with better drought-tolerance traits and richer flavor.[63]

For an overview of American wheat varieties grown in the 1800s and early 1900s, download the 1922 USDA bulletin on American wheat varieties.[64]

Signature Wheat Varieties

Throughout the decades that I have collected and grown out world heritage wheats, certain landraces stand out for their unique history, vitality, and substantial flavor. There are too many to mention here, yet I would like to share highlights of some of the most outstanding. Each is more than a plant but is a living expression of the culture, climate, and people that grew them. The children I teach call them "time machine seeds" that bring history alive in our hands, a living connection reaching across generations to us today. All of these seeds and more are available in my community seed bank, or Germplasm Resources Information Network (GRIN), for people to grow out.[65]

BLACK WINTER EMMER:
BIBLICAL ISRAEL TO NEOLITHIC EUROPE

Majestic black winter emmer stands tall against wind and weather. This rare variety sustained early peoples from ancient Israel to Old Europe. Beloved as farro in Italy, emmer withstands climate extremes better than other wheats. It is steadfast in drought, yet bears well in humid, moist weather, fortified with wild resilience against rust and fusarium. This grain, especially the hardy black winter emmer genotype, promises stable yields in the face of unprecedented climate change.

Versatile, nutty-flavored emmer is savored in soups, as a cracked grain for a hearty breakfast, cooked like rice or milled into fine flour for bread or noodles. In Ethiopian folklore it is recommended for infants, the elderly, and pregnant women for its high nutrition and digestibility. One cup of whole-grain emmer provides 24 grams of protein. Rich in fiber, protein, magnesium, and vitamins, emmer contributes to a complete protein diet when combined with legumes, making emmer grain and pastas ideal for a plant-based, high-protein food source. The Heritage Grain Conservancy emmer was collected from the Carpathian Mountains.

JALJULI BIBLICAL DURUM:
BIBLICAL ISRAEL AND LEVANT

Jaljuli wheat may be the missing link between the wild emmer wheat and the easy threshing, hull-less durums that were selected by ancient Middle Eastern farmers.

Our jaljuli was gifted by Daoud, a farmer in Za'atara in the Shepherds' Field between Jerusalem and Bethlehem. The rolling hills of Daoud's farm are abundant with diverse landrace vegetables, herbs, and grains. As we stood by his stone home, at the pre-Roman cistern that caught and stored the rainwater, Daoud quietly explained that his family had lived in this place long before the Roman exile that swept away the Jewish intelligentsia, leaving the peasant Jewish farmers on the land to grow the grain tributes for Rome.

Later when I trialed the scores of biblical landrace wheats that I collected for the Israel Plant Gene Bank, jaljuli stood out as the most robust and deep-rooted. Intrigued, I researched its history. Why was it so well adapted? I learned that jaljuli and Hourani wheats were found at Masada, stored there 2,000 years ago. Given its powerful mythic value to peasant farmers, it is no surprise that jaljuli bears an uncanny resemblance to the sacramental wheats of Old Spain, later brought to the New World.

Jaljuli (Arabic) or *gilgal* (Hebrew) means "circle." Gilgal Refaim pre-dates the pyramids and Babylonian temples by millennia. Stone circles like Stonehenge and Gilgal Refaim were sacred calendar-megaliths built by early farmers to consecrate seasonal cycles of nature and the harvest, and to invoke blessings. It is believed that Gilgal Refaim, 500 feet in diameter, was used for ceremonies honoring Tammuz and Ishtar, the gods of fertility, on the longest and shortest days of the year when the first rays of the sun shone through the northeast opening.

Concentric circles, known as "cup and ring" art, are a Neolithic design carved on stones throughout the world from India to Scotland to Peru. What do they represent? The prophet Samuel anointed Saul as king in a sacred gilgal stone circle (1 Samuel, chapters 7–11). Holy people played harps and worshiped in a gilgal (2 Kings 2:1–2).

Hourani Biblical Durum:
Biblical Israel and Levant

The Hourani durum wheat landrace was cultivated for millennia in the Houran plateau of northern Jordan and southern Syria and south into Israel. Recently, however, its meta-population has become fragmented due to the introduction of high-yielding varieties and the loss of local community seed systems. Hourani seed was collected by Nikolai Vavilov, the renowned plant explorer who combed the world for rare varieties in the 1920s. Vavilov reported:

> The Hourani landrace deserves great attention because of its exceptional quality, yield and fine appearing grain. It is distinguished by early maturity, drought resistance and resistance to lodging. It has

147

been shown in our experiments that Hourani's trait-complex is dominant in hybridization with our ordinary Russian wheats.[66]

In addition to the grains given by Daoud, Hourani and jaljuli were generously gifted by Atif and Miriam, who live in a hidden valley, down a winding path in the village of Wadi Fukin near Bethlehem. Miriam reports that the flavor is better and the breads stay fresher longer than the modern wheat varieties sold in the stores.[67]

ETHIOPIAN PURPLE: NORTH AFRICA

In Jerusalem's open-air market, Mahane Yehuda, ancient stone buildings with arched portals give way to a colorful tapestry of ethnic foods and exotic flavors. Abraham and his wife, Yehudit, opened the first Ethiopian shop in Mahane Yehuda, after they arrived in Israel from Gondar, Ethiopia, escaping local crossfire to return to their ancient homeland of Israel. Ethiopian Jews may be direct descendants of Moses's children who migrated south after the Exodus, combined with ancestors from the tribe of Dan, who fled when the Kingdom of Judah divided in the tenth century BCE, enriched with descendants from trade relations during King Solomon's time, perhaps even the Queen of Sheba.

Today, Ethiopian-Israelis number 100,000. Almost all were traditional farmers in rural mountain villages, but most have joined the ranks of Israel's low-income, underemployed populations from Third World countries. Few have found ways to adapt their farming methods to compete with high-tech farming. So they resort to shipping their ancient Ethiopian wheat, known as teff, and other traditional foods direct from Ethiopia to family-run markets such as Abraham and Yehudit's.

It was in Abraham and Yehudit's market stall that I found emmer (*Triticum dicoccum*), called *Em Ha'Hitah* or Mother Wheat in Hebrew, the almost-extinct, delicious wheat variety that was domesticated in the land of early Israel between 12,000 and 10,000 years ago.

"Do you know what this is?" I asked Abraham incredulously. "Of course, it is *aja*," replied Abraham, with an almost gleeful smile, using the Amharic word for emmer. "Abraham, this wheat was used for our first matzahs in Egypt." "Yes," concurred Abraham. "It has been kept by our people in Ethiopia." "Why don't you grow it here to bake matzahs?" "Ah," he lamented sadly. "Who of our people have farms here in holy Israel? Who would buy our simple foods?"

POLTAVKA: NEOLITHIC BLACK SEA

Landrace bread wheat was collected in 1915 in the fertile plains of Poltavka, that rise up toward the foothills of the Volga forest-steppes, north of the Black Sea in the Ukraine. Poltavka has been selected over millennia for plump, generous ears giving rich flavor.[68] The warm beauty of this landrace generates a feeling of abundance when I look at it today.

Poltavka is a site of ancient settled farming tribes where voluptuous female figurines of clay, alabaster, and marble were found among the charred wheat grains of the bread hearths, dating to before the third millennium BCE. After 2700 to 2100 BCE an Indo-European warrior culture rode in with stallions and chariots, overrunning the peaceful farming villages of old Poltavka.

ROUGE DE BORDEAUX: FRANCE

The French seedsman Vilmorin described Rouge de Bordeaux this way in 1880:

During the winter of 1870–1871 farmers in Seine-et-Oise and Seine-et-Marne, who had fled to Bordeaux because of the war, brought back a few bags of seed to sow on their land. Thus the "red wheat from Bordeaux" was introduced. Since then it has spread widely with deserved acclaim. It is productive, demanding on the soil and of a medium-tall height. Bordeaux wheat does well in rich clay soil and on a limestone subsoil it is very favorable. The grain is large, full, red and heavy. It does well spring-planted, but the maturity is a little later.[69]

The US national gene bank, GRIN, reports that Rouge de Bordeaux, released in 1884 by Vilmorin, is a selection from Noe, brought to France from Samara in southern Russia. It looks almost identical to red lammas, but has stronger stalks, greater resistance to lodging in richer fertility, and lacks the tiny awn-tail of lammas.

My Rouge de Bordeaux was generously gifted to me by Jean Francois Berthelot, French seed saver extraordinaire.

RED LAMMAS: CELTIC TO EARLY ENGLAND

Lammas is the earliest wheat sowed in colonial New England, brought over before the American Revolution by British settlers in the 1600s. Red lammas is an old Celtic wheat grown in Britain for millennia. Ancient Celts celebrated the midsummer wheat harvest in a festival called Lughnasadh (Lugh is an Irish deity whose name means Sun-King) when light dwindles after the summer solstice. Lammas (Middle English *hlaf-mas* or "loaf-mass") refers to the Celtic wheat festival where the first sheaf of wheat was ceremonially reaped, threshed, milled, and baked into a loaf. As Christianity entered, the first loaf was consecrated at a "Loaf-Mass" thanksgiving for the harvest.

There are three lammas landraces: red, yellow, and white. Red lammas is a renowned old English landrace bread wheat, described by the first modern botanist, John Ray, in 1660 in his catalogue of plants.[70] In *The Modern Husbandman* (1784), William Ellis wrote: "Red Lammas is deemed the 'King of Wheats' for having deservedly the reputation of the finest, full bodied flour. It fetches the best price at market." [71] Yellow lammas was brought by the British to the American South, where it became known as "red may," then migrated northward, becoming known as "Michigan amber." White lammas adapted well in the Northwest and became known as "Pacific bluestem." These lammas lines were an important contribution in US wheat breeding. Lammas resembles Rouge de Bordeaux in look, flavor, and baking quality, but is distinguished by a little awn-tail on the tip of the seedhead. It is better able to yield in poorer soil but tends to lodge in richer soil fertility, where Rouge de Bordeaux will stand tall, giving abundant, glowing seed. Lammas is a soft red winter wheat with aromatic flavor and crispiness enjoyed in pastries, scones, and light artisan breads. Our seed was generously contributed by Glenn Roberts of Anson Mills, and we grew it out on our farm in Massachusetts.

CAUCASUS ROUGE: TRANSCAUCASUS AND FRANCE

In the first year of my wheat trials in Canaan, Maine, I planted a field of Rouge de Bordeaux. Among these I found three unusually tall, graceful plants of extraordinary beauty. They had seedheads resembling the spring *T. carthlicum* but were winter wheats that towered above all the others. I named this new variety Canaan Rouge, and joked that it looked like an old Caucasus *T. aestivum* landrace that had crossed with *T. carthlicum*.

Years later I learned that this variety was indeed called Caucasus Rouge. A Georgian landrace bread wheat known as red doli was sown in southern France and became known as Caucasus Rouge, also rouge de roc. Its Georgian origin is well known there. Regrettably, this wheat is critically endangered in Georgia itself.

My Caucasus Rouge exhibits great disease resistance and brings a subtle fragrance of honey and nutmeg to the bread.

Rivet Blue Cone: Mediterranean and England

Gigantic, productive rivet flourishes in heavy clay soil and has a strong, sturdy stalk that supports its huge seedheads. It yields well in poor fertility and heavy clay soils where bread wheat may not.

Known to us as Poulard, Duck Bill, Rivet or Cone, this species grows tall, with a thick, rigid stem. The spikes are large, nodding as the head grows heavy with seed. One of the most esteemed of these is the Cone Wheat, so named for the conical shape of its heads, well-used for thatch due to its tall sturdy stalks. Cultivated in England and the north of France on clay-rich soils. (1840)[72]

A tetraploid closely related to emmer and durum, rivet was enjoyed throughout Europe for nourishing, high-protein frumenty pottage (thick soup or stew), a medieval staple food, and was used for ship's biscuits — a hard and dry meal. It makes excellent pasta and flatbreads that have a distinct nutty flavor. It is well documented in England since medieval times and was also used in many a thatched roof. This is an excellent species to restore as an educational project since we found it not only disease-resistant but also majestic and beautiful.

Mediterranean

A richly flavored bearded winter landrace wheat used as an all-purpose bread flour was brought to North America from Genoa, Italy, in 1819 by John Gordon. The strong naval fleet of the city-state of Genoa controlled most of the wheat trade of the Mediterranean Basin well into the mid-1800s. Wheats from Genoa at this period were brought from Cyprus, Asia Minor, and the Black Sea coast where Genoa had established colonial posts when the Byzantine Empire was overthrown in 1204. Genoa shipped wheat to ports in the south of France and Spain.

Gordon made his fortune in the wheat trade based in his home port of Wilmington, Delaware. Mediterranean wheat soon spread throughout the Eastern United States because it ripened early, escaping rust and insect damage, and it adapted well to the wet, cool weather common in the mid-Atlantic eastern seaboard. In 1862 in Mifflin County, Pennsylvania, Abraham Fultz, while passing through a field of Lancaster (also known as Mediterranean) wheat, which is an awned variety, found three spikes of awnless wheat. He sowed that seed and planted larger amounts each year until he obtained sufficient seed to distribute widely. His wheat became known as Fultz.

The Mediterranean seed sold by the Peter Henderson Seed Company in 1884 was entered into the USDA gene bank and is the variety we are restoring to commercial production from a 5-gram packet.

Kansas Turkey Red: Crimea, Russia

Crimean is a hard red winter wheat with a rich, nutty flavor from landraces that survive harsh Ukrainian winters. Turkey Red was brought from the Ukraine by Mennonites who migrated to Kansas in 1873. As Samuel Zook, bishop of the River Brethren Church, reported in 1880:

> I sowed eighty acres of Russian wheat introduced by Mennonites from Russia. We have been interviewing that field very closely all winter and have come to the conclusion that it is by far the hardiest wheat we have seen in Kansas. Our advice is to sow a larger breadth of this hardy variety next fall. If millers insist upon having the finer tender varieties let them rise up and out of their easy chairs, take to a farm and grow tender varieties themselves.[73]

Turkey Red was the hardiest wheat and soon became the favored Kansas wheat until modern breeding created shorter, higher-yielding varieties.

In the 1922 USDA Agriculture Bulletin No. 1074, the exact origin of Turkey Red is revealed.[74] Turkey Red is a Ukrainian landrace hard winter wheat from the Crimean regions encompassing the Krymka and Kharkiv areas. With this information, I have searched out Crimean landrace wheats collected by Nikolai Vavilov in the 1930s. I was able to secure a modest packet of a hundred authentic grains of Turkey Red entered in 1890 in a gene bank. I have grown them all out. The growth habits and morphology are identical, so we can restore this renowned line from its homeland parent population in Crimea.

In 1898, Mark A. Carleton was sent by the USDA on his first plant exploration trip to Russia. He brought back new durum and hard red wheat varieties mostly from Crimea to grow in the United States. Five years after the introduction of wheat from Russia, wheat production in the United States exploded, from 60,000 to 20 million bushels a year. Not only did the drought tolerance of these new varieties open up the Great Plains and the Northwest for wheat growing, the durum wheat tasted better in pasta, and the hard red wheat made better bread.

The official who sent Carleton on his 1898 collecting trip would later write, "We had forgotten how poor our bread was at the time of Carleton's trip to Russia. In truth, we were eating an almost tasteless product, ignorant of the fact that most of Europe had a better flavored bread with far higher nutritive qualities than ours."[75] The spring wheat line from Crimea, Halychanka (Red Fife), is known in Maine as Scotch red, contributing its excellent bread-baking qualities to spring wheats. Kansas's Turkey Red can be reinvigorated from the original Ukrainian Crimean collected by Vavilov and Mark A. Carleton found in the USDA gene bank.

Marquis Wheat Chew Test

In 1904, Charles Saunders developed a new variety called Marquis which soon covered the vast prairies of Canada due to its fine quality. Marquis is a cross between the early ripening Indian wheat Hard Red Calcutta and Red Fife. Hard Red Calcutta was a wheat mixture of several varieties. During the winter of 1903–1904 Saunders did not have a proper laboratory, a mill for grinding wheat, or an oven for baking bread. However, he took a few grains from each stalk, chewed them and decided on their probable flour and bread quality on the basis of the dough created in his mouth.[76]

Maine Heritage Banner

There was a time when wheat was a sure crop in Maine, raised without serious difficulty in quantities adequate to the wants of the people, the most common and reliable of crops. Wheat was a surer crop than Indian corn, more bushels of it were annually harvested and consumed. The only flour brought into our State was borne hither from vessels from Baltimore and Richmond, but seldom used except in small measure by families in our seaport towns for pastry. The Erie Canal had not yet opened. New York State's Genesee flour was unknown here.

For wheaten bread our population relied upon the home article. Few stores were without ample bins of wheat. Our wheat was not white like the brands of St. Louis. It was sweet and nutritious but of brownish hue. Much of the bran was mingled in the flour.

The best variety of winter wheat in Maine is known as the Banner wheat. In 1844 we received a small package of this grain from the Patent Office, just imported from the Baltic. We sowed it and its proceeds, till three years afterwards, a harvest of thirty-two bushels was obtained. We distributed it in various parts of the State for cultivation. As uniformly as with us, it proved a good success. Siberian, a.k.a. Java or China Tea, and Black Sea winter wheats also have been grown with great success. The Banner wheat has been raised to this day. It is a splendid grain. Sowed on grounds that the frosts do not heave badly, it is found to survive the winter nearly as well as clover. But it should be sown in August that it may get firmly rooted before winter. If a field is plowed in July, fertilized, harrowed, sown and rolled in August, or seed scattered in the cornfield previous to the last hoeing, the chance is sure of an ample harvest in July following.[77]

What was the origin of the banner wheat that did so well in Maine? Why would the US Patent Office send wheat to Maine? With this clue,

I researched the patent office activities of the period and discovered the following puzzle pieces. Henry L. Ellsworth wrote in his report of 1837:

> The introduction of a new variety of wheat promises the most gratifying results in securing production under the adverse effects of severe winters. A short time ago, the most eastern State of our Union was, in a measure, dependent on others for her bread-stuffs. That State is now becoming able to supply its own wants, and will soon have a surplus for exportation; and this is effected by the extensive introduction of new wheat. Among the varieties of this wheat, however, there is great room for selection; there is at least 20 percent difference, if regard is paid to the quality and quantity of the crop.[78]

The hardiest winter wheats originate from the Crimean regions of Kharkov, Kuban, and Samar. The very best winter wheat is Kharkov wheat — the hardiest of all known winter wheats.[79]

RED FIFE, OR HALYCHANKA: GALICIA, EASTERN EUROPE

Known as Galician spring or Halychanka in Europe, this delicious wheat has its homeland in Galicia (Halychyna in Ukrainian) in Western Ukraine. In 1842 David Fife of Ontario received from a friend in Scotland a packet of wheat from a Ukrainian ship from Danzig (Gdansk). Fife planted the grains, but only five plants germinated. Of those, some were eaten by the family cow before the last plants were saved by Mrs. Fife. Most of the wheats were winter lines, but the plants that headed in spring became known as Red Fife. Soon this cold-hardy spring wheat spread throughout Canada. Red Fife was introduced to the United States in the mid-1850s and was called in Maine Scotch Fife.

155

Vermont's Defiance

Twenty-five or thirty caged lions roam lazily to and fro, hour after hour through the day. On every side without, sentries pace their slow beat, bearing loaded muskets. Men are ranging through the grounds or hanging in synods about the doors of buildings, without purpose. Aimless is military life, except betimes its aim is deadly. The building resounds with petty talk; jokes and laughter, swearing. Some of the caged lions· read. Some sleep, and so the weary day goes by. Brattleboro, 26th, 8th month, 1863

In the early morning damp and cool we marched down off the heights of Brattleboro to take train for this place. Once in the car, the dashing young cavalry officer, who had us in charge, gave notice he had placed men through the cars, with loaded revolvers with orders to shoot any person attempting to escape, or jump from the window, and that any one would be shot if he even put his head out of the window. 28th, 8th month, 1863

How beautiful seems the world on this glorious morning by the seaside! Eastward and toward the sun, fair green isles with outlines of pure beauty are scattered over the blue bay. Though fair be the earth, it has become tainted by him who was meant to be its crowning glory. Behind me on this island are crowded vile and wicked men, the murmur of whose ribaldry riseth like the smoke and fumes of a lower world.[80]

Cyrus Pringle, Vermont plant explorer extraordinaire, amassed a vast collection of botanical samples found throughout the United States and Mexico, and exchanged seeds with scores of Europeans, with more than 20,000 samples stored at the Pringle Herbarium at the University of Vermont. More were collected for the Smithsonian and for Asa Gray of Harvard University. I have spent sweet hours poring over Pringle's eloquent descriptions of how to cross tiny flowers of wheat, the traits of grape plants, and the habits of potatoes.

In 1863, Pringle's botanical work was interrupted by the Civil War. With an abiding belief in nonviolence, Pringle, a Quaker, was imprisoned by the US military for refusing to bear arms during the Civil War in 1863. He suffered great physical hardships as a military prisoner. President Lincoln intervened and pardoned Pringle and three other Quakers. After recovering from his ordeal, Pringle continued to breed plants on his farm in Charlotte, Vermont. From 1864 to 1880, he bred wheat, oats, grapes, and potatoes.

I painstakingly searched through the USDA gene bank in 2004 and was delighted to find Pringle's wheats and oats in the collection there, waiting to be rediscovered. I excitedly mailed them to Heather Darby and Jack Lazor, as a gift for their good work to revitalize Vermont grains.

Pringle's Progress Oats

This new and distinct variety of oats was made by Mr. Pringle by crossing the Excelsior with the Chinese Hulless. In it we have a combination of good qualities which cannot fail to please — a short stiff straw and a long full head or panicle. In the Progress oat we have a head averaging as large as the largest of the taller varieties, well-filled, and being so much shorter, it does not lodge. In our trial plot of about 20 varieties of oats, the progress matured next to the Early Lackawanna oat. In the spring of 1886, from one and ¾ bushels drilled in on a rather poor soil, 162 bushels, threshers measure, were harvested. Progress oat will suit every time. Horses seem to like these oats much better than most sorts, probably because of the thin and tender shuck.[81]

Red Fife, a.k.a. Halychanka, belongs to one basic landrace with winter hardiness, resistance to drought, and excellent baking qualities. These are mostly winter wheats, but in northwestern Ukraine (Galicia) spring wheats such as Red Fife or Halychanka also thrive there. The original Red Fife or

Halychanka collected by the Vavilov Institute in Russia is available through the USDA gene bank.[82]

Banatka: Hungary, Russia, and Heritage Vermont

This renowned Hungarian landrace from Banat by the Tisza River, by the Carpathian Mountains, has exceptional winter hardiness, resistance to drought conditions, and outstanding bread-baking quality. Banatka's height towers over weeds, standing tall with strikingly attractive awned majesty. I crossed Banatka plants with the largest heads and the most tillers with the beloved Hungarian bankuti, renowned for its rich flavor. The original Banatka, famed for its adaptability, was brought to Russia in the late 1900s, was renamed Ukrainka, and soon covered vast hectares of the Ukraine. Nikolai Vavilov, the acclaimed Russian plant explorer, characterized Ukrainka as "distinguished by high productivity, excellent baking quality and broad adaptability."[83] Today my Banatka cross is the highest yielding wheat in our trials, and is being grown with good success throughout New England.

CHAPTER 5

A Taste of History

Without proper diet, medicine is of no use.
With proper diet, medicine is of no need.
— *INDIAN PROVERB*

H umans evolved from fruit-eating ape ancestors. Hominoid primates are largely vegetarians who use flat molars for grinding and lack the sharp, tearing teeth of carnivores. Charred remains from Paleolithic sites worldwide confirm that early humans ate a broad diet. Some ate more meat. Others ate more plant foods. *Diversity is the key.* Although in arctic regions only meat was available, overall the diets of our ancestors were as varied as their locations.

In Israel, 23,000-year-old hearths are strewn with thousands of charred grains as well as animal and fish bones, suggesting that wild grains were a staple food of ancient hunter-gatherers in the Levant. Hunter-gatherers ate a wide, diverse diet with a generous portion of wild grains, seeds, and tubers. Pre–Ice Age settlements from 32,000 years ago spanning Italy to Russia have grinding stones, mortars, and pestles with glossy remains of starch from wild grains such as millet, oats, emmer, cattails, and acorns. Sabine Karg, a lecturer in archaeobotany at Copenhagen University, confirms, "Carbohydrates were a key part of the paleolithic diet. Charred seeds from wild grasses are found in abundance in the pre–Ice Age hearths throughout Europe and the Mideast."[1]

Eat like a Neolithic Farmer

As the Ice Age began to melt from about 12,000 years ago, the Middle East changed from cold and dry to warm and moist. Wild edible grains, legumes, and greens thrived. Today, although the climate is drier and warmer, wild progenitor food crops still thrive in the region.

When I lived in Tekoa, near the site of an ancient Natufian early farming village south of Jerusalem, I easily collected enough wild foods to sustain me, from barley and wheat to arugula, purslane, nettle, and *chubeza* (wild mallow), and more. The wild lettuce and artichokes were too bitter and prickly to eat, but the nearby Arab village of Artas, which hosts a landrace lettuce festival each year, generously shared their seeds of a more edible strain. Almond, fig, carob, and pomegranate trees grew wild in the surrounding fields. A venerable olive tree watched over us. To my delight a pomegranate tree planted itself at my front door, probably with the help of poop from a bird, growing taller each year. Fragrant wild sage, hyssop, rosemary, and thyme were everywhere for the picking. All I needed to do to "domesticate" the wild foods was to deep-dig the soil, enriched with goat manure, and plant. Season by season the wild plants became tender and fat, replanting themselves throughout my garden, spreading like weeds. These landraces were more substantial and delicious than any organic heirloom vegetable that I've grown. I was able to gather and grow enough wild foods to eat abundantly with deep satisfaction. My next-door neighbors, Amit and Ula Biran, generously exchanged their chicken, eggs, and goat cheese for my einkorn, emmer, barley, herbs, wild greens, and heirloom seeds from Fedco. The Arab family who lived behind my house contributed *facus* (an ancient cucumber-like melon), lentils, chickpeas, pungent garlic, onions, and sheep's milk yogurt. They taught me how to make sourdough *chubez* bread on a *taboun*. I soaked almonds and landrace wheat, then blended them with dates for an almond-date-sprout milk that deeply satisfied and quenched on the hottest days. Everything was bartered, since that was more economical and gave a feeling of generosity. My little Neolithic farm community was a Garden of Eden.

THE NEOLITHIC DIET FOR TODAY

If we want to eat like a Neolithic farmer today, think about what grows on a small-scale ecological farm, not only in the cultivated fields but the nutrient-rich wild edibles in the meadow and between the vegetable rows. The wild game on my Western Massachusetts farm include pheasant, turkey, and deer, with trout and salmon from the river. Avoid commercial modern varieties, which are an empty harvest bred for appearance and shelf life. Sprout and ferment nutrient-dense landrace and heirloom whole grains; eat cheese, dairy, legumes, nuts, seeds, vegetables, fruit, fish, eggs, poultry, and as much wild food as you can forage or hunt. No processed sugar or white flour. Barter with your neighbors. Be as active as you can. Attune to the seasons.

Nectar from the Gods

Beer happens. When hunter-gatherers and early farmers stored grains, moisture occasionally seeped in, causing the grains to sprout. Sprouted grain becomes sweet due to an enzyme that converts its starch to sugar. When the sprouts are left out in the sun, they dry and become what we know today as malt. Ubiquitous wild yeasts in the air use the sugar in the sprouts to ferment when they become moist. Sprouted, dried fermented grain is not only sweet and tasty but becomes alcoholic. Who needs all the hard work of milling and baking if fermented grains create such a divine drink?

The oldest manual millstones known are trough querns. The flour is produced by the back-and-forth, rubbing-rolling motion of a stone that crushes the grain. These ancient concave querns are best suited for crushing soft, malted grain. It is difficult to produce flour from a trough quern using harder unmalted grain, which requires a flat surface to make flour. The oldest trough quern was found in Abu Hureyra, Syria, dating from around 11,500 years ago.

For at least the past 12,000 years people have eaten sourdough bread that was predigested by ever-present microbes. What was understood in ancient times as a sacred mystery still holds magic and awe for us today. A profound relationship connects each of our guts with the vital earth processes and microbial biota so that we can digest grains. It reflects the alchemical understanding of macrocosm and microcosm. Traditional fermented grain preparation practices evolved over millennia to reduce the toxicity of grain antinutrients, improve nutrient assimilation and restore probiotic digestive gut bacteria. Not only does fermentation transform our food and drinks through the life processes of microorganisms, but it deepens the flavors and aids preservation. Fermentation has been at the heart of grain cuisine since the time of the first Neolithic farmers.

The word *bread* first appears in the eleventh century CE. It means "something that has been brewed." *Hlaf*, Anglo-Saxon for loaf, is far older. It refers to rich dark soil and is related to the Slavic word *gleb*, meaning soil or earth. *Gleb* is a sound similiar to *hlaf*. Thus *bread loaf* suggests "brewed from rich soil."

HOW WAS ANCIENT BEER BREWED?

Ancient peoples may have discovered beer by tasting sprouted grain bread that had been left in a moist environment to ferment naturally. Gradually the art of producing beer evolved using sprouted grain bread sweetened with dates or honey that were fermented. Later the bread stage was skipped, and the grain was soaked in water, sprouted, dried (malted), mixed with water, sweetened with dates or honey, and left to ferment. Ancient "beer" — more like an unhopped ale or gruit — was a thin oatmeal-like porridge-brew with about 3 percent alcohol by volume.[2]

Clay tablets from Sumeria document one of the earliest beer-making recipes in "The Hymn to Ninkasi" (see page 113), the goddess of beer, a singing-instructional guide with brewing basics in a culture with few literate people. Beer is mentioned in texts on medicine, ritual, myth, and law. The Code of Hammurabi, written down in the eighteenth century BCE, discussed legal human rights, military service, trade, slavery, conduct for workers, and regulations for beer houses. Owners who overcharged customers were drowned. Priestesses found in a beer parlor were executed by fire.[3]

The Talmud[4] describes four different types of beer: *shechar* (grain beer), *pirzuma* (date beer), *t'ainy* (fig beer), and *asni* (berry beer).[5] Date beer was also enjoyed in Babylon.[6] Beer was used as a medicinal tonic to preserve herbal extracts in alcohol. Ancient North Africans brewed antibiotic beer. They stored their grain in mud bins that contained soil bacteria,

Streptomycetes, found throughout arid soils in the Middle East and North Africa. A dough was made with sprouted grain, baked briefly and then used to make a thick, fermented sour porridge. Researchers today have discovered that the North African brew contained antibiotics from *Streptomycetes*, which are used today to make the antibiotic tetracycline. Hops were added to balance the sweet malt with bitter astringency and as a preservative.

Gruit, fermented herb beverages, were enjoyed before the extensive use of hops. In the 1600s in Europe, laws enforcing the use of hops in beers may have been an attempt to regulate the considerable aphrodisiac and stimulating gruit ales by imposing the sedative effects of hops. Fermented gruit-type herb beers were well appreciated in ancient Israel. Even fermented milk, common in nomadic societies, was enjoyed. Kohanim priests who drank too much fermented milk were prohibited from serving in the temple if they were inebriated.

SPROUTED GRAIN FLOUR

Dormant wheat contains phytic acid, a.k.a. phytates, an enzyme inhibitor that protects the seed from germinating until surrounded by moisture and warmth for growth. Enzymes produced during sprouting help us to absorb the nutrients in the grains and neutralize phytic acid, an antinutrient in grains that can bind essential minerals, such as calcium, magnesium, iron, zinc, and copper. Phytates prevent the human body's enzymes from working optimally and hinder digestion. Effective methods to neutralize phytates in grain include soaking in water, sprouting, and sourdough fermentation. These bio-activating processes break down phytates and produce beneficial enzymes that enhance vitamin content and mineral availability. A biologically active, living seed is nutritionally superior to a dormant seed.

The probiotic *Lactobacillus* and other digestive microflora promoted by sprouting-related fermentation are essential for nutrient metabolism.[7] Before the introduction of modern dwarfed wheat, wheat was gathered into sheaves and left to cure in the fields, often naturally sprouting before threshing. The natural ability to sprout easily has been bred out of modern wheat.

⋙⋙⋙ *How to Sprout* ⋘⋘⋘

Cover and rinse the grain in clean, warm water. Agitate it with a spoon to skim off the lighter chaff, which will float to the surface. Place the grain in a container with sufficient water to cover it. Do not use tap water because it contains chlorine. Use spring, well, or rain water. Soak the grain for eight hours, then drain and cover. Rinse with water daily. Little white roots will emerge after about three days. Wet, germinated grain contains about 50 percent more weight in absorbed water. The original test pound should now weigh about a pound and a half. When the little white root is three-fourths to one times the length of the grain it is time to dry the grain.

Ingredients

1 cup (200 g) grain

2 cups (472 g) warm pure unchlorinated water

1 Tbsp (15 g) of an acid such as lemon juice or vinegar

Directions

Place whole grain in a bowl, float off any hulls, and cover with double the amount of water. Stir in a tablespoon of vinegar or lemon juice. Cover the bowl. Rinse daily. In 2 days the grains will begin to germinate. Rinse. Place the grain on a screen with good airflow and let it air dry, or place in a dehydrator for 14 to 18 hours until totally dry. Line trays with a nonstick sheet. Mill and bake, or store in the freezer until needed.

MALTING YOUR OWN GRAIN

We know God loves us because He gave us beer.

— BEN FRANKLIN

The growing grain sprout, or green malt, is converted to dry malt through the process of lightly drying it. To make pale malt, place a large baking pan full of sprouts over a heat source of 100° to 125°F (38° to 52°C) for 24 hours, or until the malt contains 12 percent moisture (18 ounces for the original test pound). The heat source can be a gas oven with only the pilot light on. Mix the malt every half hour to bring the bottom layer to the top, and dry it slowly. In the Middle East, the grain was put out in the sun to dry. Andrea Stanley of Valley Malt (valleymalt.com) reports, "I have not had much luck making malt in an oven. I use a dehydrator." It should be crunchy and taste sweet.

To make crystal malt, preheat the oven to 300°F (149°C) with an uncovered Dutch oven inside. When the Dutch oven internal temp reaches 300°F (149°C), place the malt in the Dutch oven. The grains will get squishy. When you can squeeze them, a sweet syrup should ooze out. Then dry the sprouts with a dehydrator or on a cookie sheet in a 212°F (100°C) oven for about one hour, until the grains turn golden brown. Crystal malt imparts sweetness and brown color to homebrew without the burnt flavor characteristic of roasted malts.[8]

Brewing Basics

Sterilize all pots and bottles thoroughly. Simmer malt (ground, dried, sprouted, hulled grain) in water for about half an hour. (This process is called "mashing.") Boil and filter the sweet, liquid, tealike wort for up to an hour. Herbs, spices, sugar, honey, or maple syrup can be added to taste. Cool the wort rapidly, then add yeast. As the wort slowly ferments at room temperature, the sugars are converted to alcohol.

KVASS "BREAD-BEER"

Kvass (Russian for "fermented") is the traditional fermented beverage of Slavic peasants, first mentioned in the Old Russian Chronicles in 989 CE, when Prince Yuri introduced Christianity to his people with a great celebration, distributing food, mead, and kvass. In *Rules for Russian Households* written at the time of Ivan the Great in the sixteenth century, the earliest recorded kvass recipe states,

Take four parts honey and strain till it is clean. Place in a jar with an ordinary loaf of bread without any additional yeast. Let ferment, then pour into a cask.[9]

There are as many recipes for kvass as there are Russian babushkas. Traditionally this mashlike mixture is made from rye bread, rye malt, or rye flour with honey or sugar and spices. Yeast may be added to increase the fermentation action. In northern Russia, ground Iceland moss or black currant leaves are added. In central Russia, caraway or even horseradish may be used, whereas in the south, sweet fruits such as pears, apples, raisins, rose hips, berries, or lemon juice is fermented. In Moscow, cranberries and raisins impart a tart-sweet flavor to the brew. Beet kvass may use a whey starter culture and later be used as a soup stock for borscht. Substitute berry juice for half the amount of water, for a substantial yet fruity flavor. Carrot kvass is considered a delicacy. Sprouted grain is often mashed or blended in as the base. Sourdough starter may be added to jump-start the fermentation culture. In the hot summer days working on my farm, we enjoy a refreshing kvass with an einkorn bread base to which I add ginger, lemon, apple cider vinegar, and maple syrup. Chill well for a tasty, nutritious thirst quencher. Garnish with fresh mint leaves.

My first experience with kvass was when I was snowed in over a long, cold winter evening on my farm with little in the larder but hard, dry einkorn sourdough bread, dried fruit, and root vegetables. Woe is me. Remembering the old stories of kvass, I soon had a pot boiling. I dried thin slices of bread in the oven till they broke like crackers. Drying caramelizes the sugars in the bread and deepens the flavor of kvass. I poured the boiling water over the toasty bread in a pot, and let it steep overnight. In the morning I strained out the sediment, added a dash of ever-present maple syrup and a handful of raisins. Wow! It was delicious hot. When it cooled, I added

Snowed-In Creativity

Einkorn chai is a delicious grain drink to enjoy hot in the long winter nights by the hearth, or chilled for a light, nourishing summer beverage. Drizzle a tablespoon of fine einkorn flour into a blender, swirling with just-boiled water. Lightly simmer and stir for a half hour or so. Add a dash of maple syrup. Garnish with nutmeg. Optional: freshly grated ginger tea essence and/or vanilla.

sourdough starter, which speeded the fermentation. For days afterward I delighted in the beverage, especially when it began to ferment into a tangy, earthy flavor. It was a hit when neighbors came over for a potluck evening of storytelling and song by the hearth.

Kvass is a lightly fermented, slightly alcoholic beer. Homemade kvass is made nowadays from malted rye, barley, or wheat and rye, wheat, or buckwheat flour or from pastry, bread, or rusks; with honey, various fruits, berries, or herbs to flavor it. A twelfth-century manuscript mentions "much drink and Kvass." It is this that seems to be referred to by Tedaldi, even though the drink he mentioned in 1581 was made with "water mashed with oatmeal, then cooked, as the general drink. Oats relieves the badness of the water and makes the men fat."[10]

A traditional kvass recipe:

To make kvass put a pailful of water into an earthen vessel into which one shakes two pounds of barley or rye meal, salt and honey according to the wealth of the household. In the evening place in the oven with a moderate fire and stir occasionally. In the morning, pour the clear liquid off to ferment and enjoy.

Eight quarts water with 1½ lbs malt, 1 lb. rye flour, 1½ lb. honey, ⅛ of a lb. mint leaves, half pepper pod, and half cake of yeast. Mix the malt and flour with boiling water and make a thick dough. Put into barely warm oven, and leave for the night. Next day dilute dough with eight quarts boiling water and pour into a wooden tub. Let stand for 12 hours, then sieve through a cloth. Pour one quart into an enamel saucepan, put on fire, add 1½ lb. sugar, and an infusion made with the mint leaves like weak tea. Boil once, then take off fire, cool until just warm. Add the yeast previously diluted with one cup of this same warm liquid. Let stand in warm place until it begins to ferment; then pour it into the rest of the kvass in the wooden tub, and let stand until bubbles appear. Prepare clean bottles, putting raisins into each; pour in the kvass, cork the bottles, tie the corks with string to the necks of the bottles, and keep in a warm place for a day or two. Then put in a cold cellar.[11]

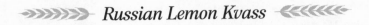 ## *Russian Lemon Kvass*

Ingredients

3 organic lemons, sliced thinly and seeds removed

¾ cup sugar (150 g), maple syrup (238 g), or honey (255 g)

5 tsp (14 g) active dry or beer yeast

6 Tbsp (74 g) raisins

Directions

Place sliced lemons in a 5-quart (4.73 L) container and cover with 4½ quarts (4.26 L) boiling water. Add sugar, maple syrup, or honey, and let cool. Add the yeast and raisins, and ferment in warm place for a day. Strain into bottles, and refrigerate.

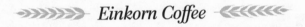 ## *Einkorn Coffee*

This drink imparts a harmonious feeling of energy that enlivens without the nervousness energy of coffee alone.

Ingredients

Organic ground coffee

Ground roasted sprouted einkorn (malt) or ground roasted
    einkorn grains

Roasted burdock root, chicory root, dandelion root

Optional: chai-type spices such as nutmeg, cardamom,
    cinnamon, or cloves

Directions

Mix all the dry ingredients together. Boil, then simmer for 20 minutes. Strain. Add milk or sweetener to taste, and enjoy.

 *Grain Berry Kvass*

Ingredients

1 cup (150 g) whole raw cranberries

1 Tbsp (15 g) einkorn flour

1 Tbsp (15 g) grated lemon peel

1 Tbsp maple syrup (20 g) or honey (21 g) to taste

Your preferred fine white or red dessert wine

Directions

Place cranberries, einkorn flour, lemon peel, and a dash of maple syrup or honey in 2 quarts (2 L) of boiling water. Simmer for half an hour and strain. Add the wine to taste when cool.

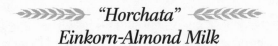 *"Horchata"*
Einkorn-Almond Milk

I lived on refreshing icy-creamy homemade einkorn-almond milk when I worked in the scorching heat of the Middle Eastern fields collecting seeds. Traditional Horchata, made with almonds and barley, has been enjoyed since ancient times in the Middle East and was brought to Europe by Arab invaders from the eighth to the thirteenth century. It is known as a medicinal energy drink with digestive benefits.

Ingredients

2 cups (472 g) raw almonds

½ cup (100 g) einkorn grains

6 diced dates

Dash of nutmeg to taste

Scant tsp (5 g) grated ginger

1 tsp (5 g) vanilla extract

Cinnamon sticks for garnish

Directions

Soak the almonds and einkorn grain in water overnight. Add dates, a dash of nutmeg, ginger, and vanilla to the einkorn-almond milk and blend. Strain. Chill. Blend right before serving for a light froth. Garnish with a cinnamon stick and serve.

 *Einkorn Milk*

Ingredients

2 cups (480 g) boiling water
1 Tbsp (15 g) sifted einkorn flour
Dash of nutmeg, ginger, and maple syrup to taste
Vanilla extract to taste (optional)

Directions

Boil the water, and carefully pour into the blender. As it is blending, drizzle in a heaping tablespoon of einkorn flour. Add a bit more flour if you prefer a thicker beverage. Simmer on stove for 15 minutes, stirring well from the bottom of the pot. Add nutmeg, ginger, and maple syrup to taste. Add vanilla, if using. Delightful hot or cold.

SPROUTED BULGUR

When I first experimented with dehulling einkorn by hand, I stumbled on an ancient method for hulled grain preparation: bulgur. The traditional method to remove the hulls from the grain is to pound and abrade the grains with a mortar and pestle, or even in a hollowed-out tree trunk. After the pounding, the light hulls need to be separated from the heavy grain. The easiest method to remove the hulls is to put the grain in water and float off the hulls. I swished the grains thoroughly to remove the hulls. By the time all the hulls floated off after several rinsings, the grains had begun to absorb water. What could I do? Lacking a sun-drenched rock on which to spread out the grains to dry, I spread them out on a towel. They began to sprout before they were totally dry. The next day I placed the grain in a warm 200°F (93°C) oven for three hours to make sure they were fully dried.

Without realizing it, I had re-created the traditional Middle Eastern dish of sprouted bulgur.

Nutty-flavored sprouted bulgur may be the most ancient food prepared from hulled wheats. Modern store-bought bulgur is cooked, then dried, but the critical stage of sprouting is no longer done. The original method of sprouting the grain before eating is not only more delicious than today's boiled, cracked bulgur, but it is a living food rich in vital enzymes that enhance digestion. Known as *siyez* in Turkey, einkorn's region of origin, the preparation of bulgur is a festive community celebration of the harvest. Women gather in their homes to pound, soak, and simmer the kernels till they swell. During the simmering process, the outer bran, rich with nutrients, dissolves in the water, then is reabsorbed into the grain. The grain is then spread out on large cloths and dried on sun-drenched roofs. The simmering and sun-drying protect the kernels from germs and insect eggs for a clean, pathogen-free grain product with a long shelf life. Bulgur is a traditional fast food since it needs only a brief soak in warm water to be eaten.

Frumenty Fermented Grain Porridge

Echoing Middle Eastern bulgur, frumenty, from the Latin *frumentum*, meaning "grain," is the boiled cracked-wheat dish enjoyed in Europe from Neolithic times through the Middle Ages until potatoes were introduced. Frumenty was the staple food, perhaps more than bread, for European peasants and is enjoyed in rustic cuisine to this day, as I discovered during a delicious heritage food festival in Tuscany, Italy. A thick, flavorful frumenty porridge was served as a base for soup, salads, and meat and even as a dessert garnished with cheese drizzled with honey. It can be enjoyed as a breakfast cereal with raisins, cinnamon, and milk.

Frumenty Recipes

The typical method of preparation was to boil whole grains of wheat in water or milk, sweeten with honey, and flavor with spices such as cinnamon. Here is a traditional recipe that is so old that it is in Middle English: "To make frumente. Tak clene whete and braye it wel in a morter tyl the hulles gon of; seethe it til it breste in water. Nym it up & lat it cole. Tak good broth and swete mylk of kyn or of almand and tempere it therwith. Nym yelkes of eyren rawe and saffroun and cast therto; salt it: lat it naught

171

boyle after the etren ben cast therinne. Messe it forth with venesoun or with fat moutoun fresch."[12]

In other words: Take clean wheat and pound it in a mortar well so that all the hulls come all off. Seethe (boil) it till it burst, take it up out of the water, and let it cool. Take fair fresh broth and sweet milk of almonds, or sweet milk of kine (cow's milk) and mix it all together. Add egg yolks. Simmer. Serve with fat venison or fresh mutton.

Country women in shawls and sunbonnets used to come to the market with little carts carrying basins of new wheat boiled to a jelly and lightly cooked with milk, egg, and raisins. The mixture was poured into pie dishes and served on Lent Sunday and during the following week. A "healthy" dose of spirits is often mentioned as accompanying frumenty.[13]

Kolio: Caucasus Frumenty

A fermented grain dish, *kolio* traces its documented history to the fourth century CE but is likely to have been passed down from early farmers. Of twenty species of wheat known today, twelve are indigenous to Georgia. Georgian traditions of grain preparation reach far back into the distant past. Ingredients: wheat, honey, walnuts, raisins, and water.

It takes three days to make kolio. One starts by cleaning the wheat, floating off the hulls, rinsing it in water, and briefly warming the grain in a large heavy pot of water. The pot is taken off the fire and covered with thick blankets for a day so that the wheat can slowly absorb the warm water. When the blankets are removed the next day, the pot is warm and fermenting. Honey, chopped walnuts, and raisins are added, and the dish is stirred to bring in air until the flavor is well developed, in about three days.

FRIKAH ROASTED SPRING WHEAT

Parched green wheat, known as *frikah* in Arabic, *kali* in Hebrew, and *Grunken* in German, is one of the oldest methods to eat landrace wheat known to humankind. It has been prepared in the Middle East for millennia. After long winter months of waiting for the new harvest, hungry farmers reaped the wheat in its first edible stage. The fresh spring wheat is lightly roasted in an open fire when it is almost ripe but still chewy green. It has a sweet-savory, smoky taste. Parched grain was a staple food in biblical times, and spring wheat was offered in the temple in Jerusalem.

In Old Europe as well, due to the tight hull encasing einkorn, emmer, and spelt, early farmers devised methods to remove the hull encasing the mature hard grains, such as pounding with a mortar and pestle. Soaking the sheaves in water, then roasting them over open flames is the easiest method.

The book of Ruth recounts how roasted spring wheat was eaten during the barley harvest because it was the only grain available before the mature wheat harvest. In the first and second books of Samuel, roasted wheat is combined with other dried foods such as nuts and raisins that are easily carried on journeys. Mention of roasted spring wheat is found in the following biblical texts:

If you bring a grain offering of the first fruits to the Lord, offer the crushed heads of the spring aviv grain roasted in the fire.[14]

And ye shall eat neither bread, nor parched grain, nor green ears, until the selfsame day that ye have brought an offering unto your God: it shall be a statute forever throughout your generations in all your dwellings.[15]

The day after the Passover, the very day they ate of the produce of the Land, the unleavened bread and the parched grain.[16]

When she gleaned with the harvesters, Boaz offered her parched grain. Ruth ate parched wheat in the field with Boaz. She ate all she wanted and had some left over.[17]

Jesse sent young David to bring food to his brothers who were fighting the Philistines. David brought parched grain and bread.[18]

The Talmud explains the methods for parching spring wheat for the temple offering:

They reaped the spring wheat, put it into baskets and brought it to the Temple in Jerusalem. Then they parched the green wheat with fire in order to fulfill the precept that it should be parched with fire. The Sages explain that fresh and tender grains were beaten with reeds so the grain was not crushed. Then they put it into a long pot perforated with holes so that the fire would take hold of all of the grains. They then spread out the parched grains in the Temple Courtyard so that the wind would blow over it to cool and dry it. Then they put it into a special gristmill that carefully separated off the husks without

173

damaging the tender grain. They removed a tenth of an ephah. What was left over was redeemed and might be eaten by anyone. Rabbi Akiva declared it was liable to the dough offering and to tithes for the poor and hungry.[19]

The stage at which frikah is harvested is critical. Like harvesting sweet corn for roasting, there is an optimal time to harvest wheat for roasting. When the leaves just begin to turn slightly yellow but the kernels are still soft and creamy is the ideal moment. If the wheat is immature, the grain will collapse. If it is too mature, the grains are hard and doughy. The right time for harvest is between the "milk stage," when the soft endosperm can be still be squeezed out from the grain, and the "dough stage."

The harvested green wheat sheaves are then placed out in the sun to dry for several days. The brown seedheads are dipped in water, then the sheaves are either held over an open flame or placed over piles of dry brush and set on fire. The blaze must be carefully controlled so only the straw and chaff burn, not the seeds. The chaff is just charred. The high moisture content of the milky kernels prevents them from catching fire. Close attention must be given to the fire, the wind, and the progress of the burn to ensure a good, lightly roasted final product. The roasted wheat then is cooled in the wind. The seedheads are left out to cool, then are threshed or rubbed off and sun-dried. It is this threshing or rubbing process of the grains that gives its Arabic name, *farīk*, meaning "rubbed" and called "*frikeh.*" The seeds may be cracked into smaller pieces so they look like a green bulgur or left whole to be cooked like rice. Frikah is traditionally enjoyed for stuffing squash, eggplant, and grape leaves; boiling in soups; or as a bed for other dishes like chicken. Roasting spring grains converts the starch into a sweet smoked sugar. The resulting food is earthy and smoky with a distinct flavor. Cooked like rice, parched spring wheat has a higher nutritional value than brown rice.

Bread from the Earth:
The Simplicity of Sourdough

Sourdough happens. Ubiquitous microorganisms float all through the air, swim in the water, and live in our gut to help us digest food. Microbes love einkorn. Its nutrient-rich digestibility creates a vital substrate for beneficial microbes to thrive. From the moment that water comes into contact with flour, the wild yeast and lactic-acid microbes that give starter its leavening properties begin to grow, fermenting flour and water into sourdough. The microbes consume the sugars in the flour, breathing in oxygen and breathing out the carbon dioxide that causes dough to rise when baked. Sourdough microorganisms introduce a rich complexity of flavor with a subtle, aromatic tang that imparts superior flavor and nutrient absorbability to bread. The acidity in sourdough helps bread stay fresher longer. The fermentation process strengthens the adhesion of the gluten matrix, giving a better crumb structure.

Making a sourdough starter is an easy natural process of combining einkorn flour with pure water. I regard my sourdough microbes as tiny "pets" to care for. The glass jar in my fridge contains the bubbly, thick, pancake batter–like mother starter. I feed my "pets," that is, refresh the starter, by adding equal amounts of spring water and flour every few days and especially the night before I plan to bake. (See the preferment section on page 176 and appendix 1 for details.) After taking out a portion for my bread dough, I mix more water and flour into the remains at the bottom of the starter jar, perking it up with renewed bubbly life. If I go away for a long trip, I may clean out the jar and start over if the smell is too acrid upon return. Since antiquity a portion of unbaked dough was saved as a starter for new breads. Pliny the Elder reported that the Gauls and Iberians used beer foam to bake "a lighter kind of bread than other peoples." Wine-drinking cultures such as ancient Israel used a paste of fermented grape skins and flour or wheat bran steeped in wine as a source for starter.

HOW TO MAKE EINKORN SOURDOUGH STARTER

About a week before you plan to bake, mix equal amounts of flour and pure water; that is, well water, spring water or distilled water. (Do not use chlorinated water. It will destroy the beneficial microbes.) Repeat the feeding each day for about seven days. After a week of feeding, the starter should be mature, active, and ready to use. You will observe that freshly fed starter will rise and bubble through the day, then later collapse when the sugars are fully metabolized. To create an active starter uniquely adapted to your local water, flour, and temperature, observe the timing of this cycle in your starter. The best time to feed the starter is when your starter has risen up and just slightly begins to pull down. This is the ripe moment when the microbes are active and hungry.

Day 1

Mix together equal amounts of water and einkorn flour. For example, 2 Tbsp (30 g) warm water and 2 Tbsp (30 g) einkorn flour. Mix well, and place in a glass container with a loose top to allow airflow. Stir a few times a day if possible. Store in a cool dark place or in your refrigerator in the summer.

Day 2 thru Day 5

Each day stir in 2 Tbsp (30 g) warm water and 2 Tbsp (30 g) einkorn flour. Your starter is mature when there is a lovely aromatic sweet tangy fragrance and small fermentation bubbles arising throughout the mixture. It may take several experimental batches before you develop the best sourdough adaptation for your unique conditions. Be patient, persevere, and keep at it!

Preferment

Prefermentation is a process to feed the microbes in the sourdough starter by adding more water and flour each day before mixing the bread dough, so that the starter is highly active and ready to use. It is especially important to feed your mother starter the day before you plan to make sourdough bread to get a good rise. This is an age-old practice throughout Europe and the Middle East. Each country has unique preferment traditions. In France, a preferment is called *levain*. It is a *biga* in Italy; in Germany, *sauerteig*. Originally from Poland, a *poolish* preferment is known as a sponge in the United States. The subtle variables in each sourdough process such as temperature, water quality, and the air itself will give your sourdough a unique terroir all its own.

Discover for Yourself

I must be frank: when I read cookbooks, I look for the overall proportions and often adapt recipes by feel. I learned baking from grandmothers and from playing in the kitchen. So the best way to understand baking with einkorn is to experiment on your own in small batches and compare proportions and ingredients. What is the basic ratio of flour to water? Proportions range from 3½ cups flour to 1 cup water to 3 cups flour to 1½ cups water, which is good for slow-rise Dutch oven bread, to 5 cups flour to 1½ to 2 cups water. The effects of temperature? Timing? Sourdough or yeast? Tap, rain, or spring water? Overnight slow-rise in the fridge? How does sifting out the bran affect the water absorption rate and ratio? Experiment with substitutions, for example: using potato water or cream for the liquid, adding an egg, or grating in cheese. Keep careful records to replicate your successes. Let the mystery of baking transform into experience in your hands. May delicious creativity soar!

Einkorn Baking Tips

Soak grains overnight to activate beneficial enzymes and break down antinutrient phytates.

Wetter dough. Einkorn absorbs liquids and fats more slowly than modern wheat. The dough is wetter at first. Wait fifteen minutes after mixing the ingredients before you fold or knead so the liquids and fats are well absorbed. If you add more einkorn flour to decrease stickiness, the bread may be too dense later. An overnight slow-rise helps butter or oil incorporate fully. Oil or wet your hands to fold the sticky dough. Use a scraper to shape dough.

Yeast and rising. Einkorn's delicate gluten cannot support the profuse yeasty bubbles of most modern wheat bread recipes. Decrease the amount of yeast by up to half in recipes for modern wheat. Mix, let rest, then shape your loaf. Let slow-rise in a cool place. All recipes herein use instant (a.k.a. rapid rise) yeast that can be mixed directly into the flour and does not need to be preactivated in the water. Another tip to enhance lightness is to rigorously beat whole eggs on high speed in a mixer, then fold in.

Slow-rise overnight in a cool place. For yeasted breads, mix the dough in the morning, let it rise slowly at a cool temperature, and bake after four to six hours, or better yet — overnight in the refrigerator. Even if you use yeast instead of sourdough, an overnight slow-rise in the refrigerator produces the fullest flavor and gluten adhesion.

Use sourdough. Bio-enlivened flour is easier to digest, enhancing beneficial enzymes and nutrients, has a richer flavor, and is healthier.
Always refresh sourdough a day prior to using in a preferment.

No kneading. Einkorn bread doughs need minimal kneading. Gentle
folding works best. Excessive kneading does not increase einkorn's
gluten. Time does. Cover dough with plastic, and slow-rise in a
cool, dark place so that the gluten matrix can knit together. An
overnight slow-rise in the refrigerator is the best.

No punch down. Einkorn's gluten is delicate. Mix, fold, or shape it,
slow-rise ferment in a cool place (overnight if possible), and bake.
A second rise is not necessary.

No preheat. To optimize rising, do not preheat the oven. Put the bread
in a cold oven, then turn on the heat.

Less liquid and fat. Einkorn may be substituted for whole-wheat flour
in recipes; however, decrease the amount of liquid and fat by 5
to 10 percent. Whole-grain flour absorbs more water than sifted
flour. Be patient.

Weight to volume. When you mill your own flour, you will produce
a greater volume of flour from the denser grains. For example: 3
cups (600 g) grain makes 5 cups (600 g) flour.

Blessings are hidden. The carotenoids in einkorn dough will oxidize and
darken when exposed to light. Ultraviolet rays in sunshine cleanse
and sterilize; therefore, fermentation is traditionally protected
from sunlight. Store fermenting einkorn dough in cool darkness.

Storage. Avoid storing bread in the refrigerator where it will dry out.
Store in a paper or cloth bag at room temperature.

Bread makes itself, by your kindness, with your help, with imagi-
nation streaming through you, with dough under hand, you are
breadmaking itself. This is why breadmaking is extraordinarily
fulfilling and rewarding. Love is not only the most important
ingredient, it is the only ingredient that really matters.

— Edward Brown, *Tassajara Bread Book*[20]

Einkorn Kvass sweetened with raisins, recipe for Grain Berry Kvass on page 169.

No-Knead Artisan Einkorn Bread, recipe on page 179.

9

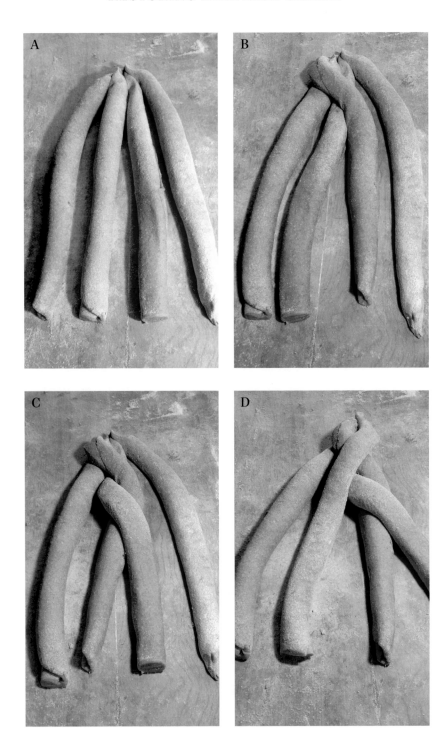

Four-Braid Challah: (A) Divide dough into four equal pieces. Roll each section flat, then tightly roll up each section like a rug. Line up the sections and pinch the ends together. (B) Take the left strand and move it over to the right two strands. (C) Then move the same strand under the strand to the left. (D) Take the far right strand and move it left over two strands. Then to the right under one strand. (E) As you braid say to yourself: Left side, over two and under one. Right side, over two and under one. (F) Repeat steps until the strands are braided, then pinch the ends together and tuck under. The finished loaf before baking. (G) The baked loaf.

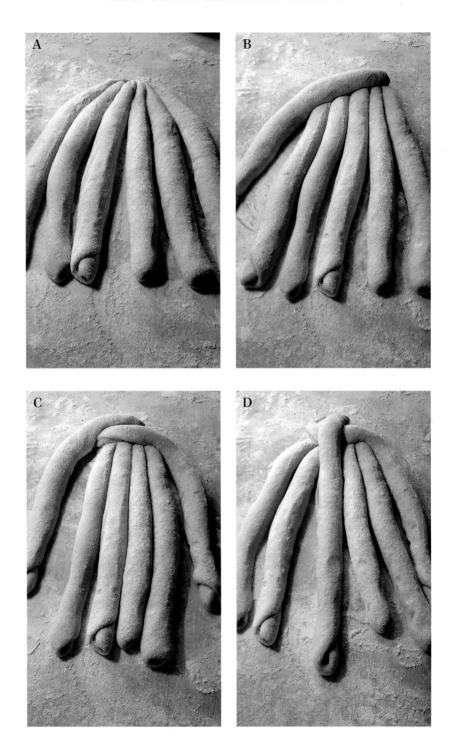

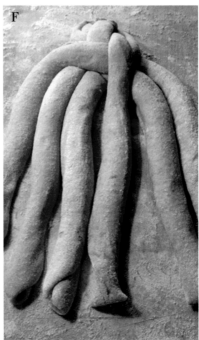

Six-Braid Challah: (A) Place the six strands in a row parallel to one another. Pinch together at the top. (B) Move the outside right strand all the way over to the left. (C) Take the second strand from the left and move it to the far right. (D) Take the outside left strand and move it two strands to the right. (E) Move the second strand from the right over to the far left. (F) Take the outside right strand and move it two strands to the left. (G) As you braid say to yourself (alternating sides): "Second up. End to the middle. Second up. End to the middle." When all the strands are braided, pinch the ends together and tuck under.

Einkorn Bagels, recipe on page 191.

Pizza, recipe on page 194.

Einkorn Kreplach, recipe on page 197.

Sculpty Dough for Pastry Art, recipe on page 205.

Pastiera Ricotta Cheesecake, recipe on page 206.

Wine Cream Biscuits, recipe on page 213.

⋙ No-Knead Artisan ⋘ Einkorn Bread Baked in a Dutch Oven

This is a basic traditional sourdough recipe that can be a foundation for creativity. It is incredibly easy to make even if you've never baked bread before. It looks so beautiful, no one will believe you're not an experienced baker. Refer to the "Bread from the Earth" section on page 174 for details on preparing the Mother Dough (preferment).

Ingredients

MOTHER DOUGH (PREFERMENT)

2 Tbsp (30 g) activated sourdough starter or ¾ tsp (2 g) yeast

½ cup (120 g) warm pure (unchlorinated) water

1 cup (120 g) einkorn flour

DOUGH

5 cups (600 g) unsifted whole einkorn flour

1 tsp salt (6 g)

1¼ cup (300 g) warm pure water

1 Tbsp (15 g) olive oil

1 Tbsp honey (21 g) or maple syrup (20 g)

All of the Mother Dough (268 g)

½ tsp (1.4 g) dry yeast or ¼ cup (60 g) sourdough starter, if not using the
    Mother Dough as a starter

Directions

Mix the flour and salt together. Mix the warm water, oil, and sweetener together. If you are not using the Mother Dough as a starter, add yeast to the dry ingredients or sourdough to wet ingredients. Mix all the ingredients together. Let the dough rest for 15 minutes so that the liquids are well absorbed. This is a wet dough, so do not be tempted to add more flour. On a well-floured work surface, use a dough scraper or your oiled hands to fold dough to the center several times to create a ball-like shape. Place on parchment paper in a large bowl. Cover with a plastic bag or wrap. Let it rise slowly in the cool darkness of a refrigerator overnight.

The next day, preheat an oven-safe heavy pot for 30 minutes at 450°F (232°C). Carefully remove the hot pot from the oven with mitts. Use parchment paper to lift the dough into the hot pot. With the dough and parchment paper in the pot, cover and bake for about 45 minutes. For a crustier loaf, bake for 5 more minutes in uncovered pot. Cool and enjoy.

Variations: Crack an egg into the measuring cup before adding the warm water (still measure up to 1¼ cups). Grate ½ cup (118 g) cheese into flour. Substitute a tablespoon of cream for oil and/or warm milk for water at a 1:1 ratio. Add a cup of blended einkorn sprouts. (Blending avoids hard sprouts on the crust.) For enhanced moistness, substitute strained drained potato cooking water for plain water.

 *Challah*

Bread was offered to the Hebrew Goddess on a golden table in the desert tabernacle.[21] My home-baked challah sits on a simple hand-carved Shabbat breadboard, but it certainly does have a heavenly aroma. Eastern European challah resembles an eggy brioche that is crusty and braided. With more delicate gluten than modern wheat, using less liquid, more eggs, and salt helps einkorn challah dough hold its shape. A teaspoon of vital wheat gluten helps einkorn challah keep the braid well shaped. It looks so beautiful when the dough is twisted into a traditional braid. The secret to rich challah luster is two brushings of egg wash.

The meaning of the challah in ancient Israel was the dough portion of bread that was gifted to Levite musicians and Kohanim priests. I separate a portion of each dough and burn or compost it with a silent blessing to the earth, or bake an extra dough to gift to a friend. This is an easy, foolproof traditional challah recipe passed down to me by my Nana. Makes 2 loaves.

Ingredients

5 medium eggs, plus 1 for egg wash
¼ cup maple syrup (80 g) or honey (85 g)
¼ cup (60 g) olive oil, melted butter, heavy cream, or yogurt
1 Tbsp (15 g) vanilla extract

1 tsp (3 g) yeast or ¼ cup (60 g) refreshed sourdough starter with
  1 Tbsp (15 g) more flour for the dough
4 cups (480 g) sifted einkorn flour (plus ⅓ cup [40 g] for kneading surface)
1 tsp salt (6 g)
1 Tbsp (12 g) vital gluten to hold the dough together better.
1 Tbsp (10 g) sesame, chia, or poppy seeds to sprinkle on the
  braided dough (optional)

Directions

Whisk or blend together 5 eggs, maple syrup, olive oil (melted butter or cream) and vanilla extract. Add sourdough starter to wet ingredients or yeast to flour. Mix together the flour, salt, and vital gluten (and yeast if using). Mix wet ingredients into dry ingredients. Let dough rest at least 15 minutes so that the liquid ingredients are well absorbed into the einkorn flour. Dust working surface generously with flour. Knead and fold until smooth.

Note: Dough consistency depends on complex variables such as flour grade, the size of the eggs, and even the humidity in the air. Be ready to adjust by slathering the working surface with oil if your dough feels too dry, or add a bit more flour if it is too moist. Einkorn flour's gluten is too delicate to hold the braids together unless you add vital wheat gluten for tighter braids.

Traditional Blessing: Place your hands over the dough and say a silent blessing, then remove a small portion. Place it in your compost to return to the earth, or burn in the oven as an offering. At the table, place the fragrant baked challah in the center, invite everyone to place hands on the challah, chant a blessing, and break bread.

Braiding: The dough should be tight so the strands hold their shape well. Divide the dough into even pieces according to the number of strands you will braid. For well-formed braids, compress each section into a flat oval, then roll it up tight like a rug. See the pages 10 to 13 in the color insert for photographs and instructions on how to braid 4- and 6-braid challahs.

To make the 6-braid challah, divide into 6 equal balls. Flatten and roll up. Place floured parchment paper on a large baking sheet. Place the 6 strands in a row parallel to one another. Pinch together at the top. Move the outside right strand all the way over to the left. Take the second strand from the left and move it to the far right. Take the outside left strand and move it to the middle. Move second strand from the right over to the far left. Continue until all strands are braided. As you braid say to yourself (alternating sides): "Second up. End to the middle. Second up. End to the middle." Cover with plastic wrap. Refrigerate overnight. The next day, beat one egg with 1 tsp (5 g) cold water; brush the egg wash on the loaves. Optional: sprinkle with sesame, chia, or poppy seeds. Preheat the oven to 375°F (191°C). Brush again. Bake for about 40 minutes till golden.

⟫⟫⟫⟫⟫ *Babka and Pashka* ⟪⟪⟪⟪⟪

A *babka* is a grandmother or old woman and the traditional eastern Europe festive braided bread baked in spring since Neolithic times by the matrilineal early farmers of Old Europe over 7,000 years ago. It certainly does look like challah. Or does the challah look like a babka? The babka represents the ancestor grain mother and birth-giver. *Pashka,* believed to originally be from *pesach,* the Hebrew word for Passover, in eastern Europe refers to the Easter celebration of resurrection and male fertility, an unusually tall, erect bread rounded and swollen at its top. Ahmm. Yes, you guessed it.

The baking of the babka and pashka is done with millennia-old rituals. Baking pashka was a central event of the year for the woman of the house. Even the ashes remaining from the oven after baking the pashka were scattered over the garden when the first seedlings were planted. The eggshells were hung on tree branches as a celebration of spring renewal. The finest wheat flour and abundant eggs are used for the Easter pashka and babka bread.

Traditionally there were three types of babkas or pashkas, each baked symbolically: the yellow babka (with added egg yolks) for the sun and sky; the white babka (with fine flour) for the departed ancestors or the air to bring blessings; and the black babka (with added rye flour) for the family or the fertile earth, flavored with spices and roots. Each was baked on a different day with special preparations. When the dough was rising, all other adults had to leave the house, and the children were sternly instructed to remain silent so as not to disturb the holy bread. If the pashka did not rise well or a babka baked unevenly, it was an omen of family misfortune for the coming year.

Directions

Use the basic challah recipe for these sublime festive breads. Options: Substitute 1 cup (122 g) of tapioca flour for 1 cup (120 g) of sifted einkorn, and heavy cream in place of olive oil for a richer bread. To bake the pashka, oil and flour the bottom of a tall can. Roll a cylinder of well-kneaded dough into a rectangle of parchment paper. Insert into the can. Create a decorative braided topping for the pashka and place it on top of the can.

Grandmothers' Babka Prayers

For the yellow pashka with many egg yolks:

Holy Pashka, may you be as great and beautiful as the sun, because it is for the sun that we bake you. May all (family members mentioned individually) who are alive be healthy. May our children grow as quickly and finely as you are. Shine for us, Pashka, as the holy sun shines. May our bread in the field be as rich and great as you are.

The "white" pashka heard:

May the righteous souls be as pure and holy as pure, holy, and great is this pashka. May the souls be as happy and comfortable as the pashka in the oven. We are baking this pashka for you, our ancestors, we are honoring you, and in turn, may you help us. May your time in paradise be as beautiful as these pashky in the oven.

In placing the "black" pashka in the oven, the woman expressed honor and respect for mother earth, and wished people and all farm animals health and well-being. She prayed for a bountiful harvest, and for no storms, lightning or hail.[22]

⋙ *Hrudka Egg Cheese* ⋘

Hrudka is an eggy ricotta-type cheese traditionally eaten in or with a pashka. It is made from eggs and milk, infused either with blood-red beets and horseradish or sweetly spiced with nutmeg, cinnamon, and vanilla and a generous spoonful of honey.

Ingredients

12 eggs (600 g)

1 quart (0.95 L) whole milk

¼ cup maple syrup (80 g) or honey (85 g)

½ tsp salt (3 g)

1 Tbsp (15 g) lemon juice

1 Tbsp (15 g) grated beets and horseradish, or a dash of cinnamon, nutmeg and

1 Tbsp (15 g) vanilla extract

Directions

Whisk ingredients together and simmer in a heavy pot, stirring from the bottom till the curds form. Drain well in a colander, and refrigerate. Traditionally enjoyed on the top or inside of pashka or babka Easter breads.

⋙ *Celebration Bread* ⋘

Celebration breads are a forgotten art being rediscovered today by artisan bakers. In Europe, where wheat has been grown for millennia, festive breads are intricately decorated with dough shapes of the sun, moon, birds, plants, and animals. Their origin is in the ancient pagan belief in the mystical, life-giving fertility of wheat. The Ukrainian word for grain, *zbizhzhia*, means "the totality of divinity." The sacredness of bread is used to evoke blessings at a marriage or birth of a child. Special breads are buried with the departed. Ornate decorative breads are placed on the altars in church at harvest time. Throughout rural Europe, wedding breads are made with age-old rituals. Women prepare the bread while singing traditional wedding songs. The bride and groom are blessed with it before the marriage ceremony; then the bread is offered to the guests. People ritually break richly decorated traditional breads together. This may be an early expression of collective communion. Today the ritual breads may have lost their pagan connection but are vital in folk traditions and echoed in church ritual.

There are two lists of ingredients; one for salt dough, which is inedible and made to be saved, and one for art dough, which is meant to be eaten and should be stored at room temperature. The directions are the same for both sets of ingredients.

Ingredients

SALT DOUGH

2 cups (240 g) flour

1 cup (273 g) salt

1 Tbsp (15 g) cream of tartar

2 Tbsp (30 g) oil

2 cups (480 g) warm water

1 egg white, for glaze

ART DOUGH

2 cups (240 g) flour

1 cup (200 g) sugar

1 Tbsp (5 g) cream of tartar

1 Tbsp (15 g) oil

2 cups (480 g) warm water

1 egg white, for glaze

Directions

Mix together the flour, salt or sugar, cream of tartar, and oil. Slowly add warm water, and knead until soft and pliable like play dough. Form shapes. Dry at 300°F (149°C) on a baking sheet lined with parchment paper. Immediately after baking, insert toothpicks to hold the shapes on bread. Glaze with egg white.

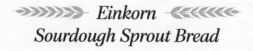

Einkorn
Sourdough Sprout Bread

This is my favorite recipe of all. It reminds me of the vital, chewy whole-grain Bavarian breads that I enjoyed in Europe — with substantial flavor yet easily sliced for sandwiches.

Ingredients

½ cup (100 g) uncooked einkorn berries

1½ cup (360 g) spring water for sprouts (plus extra for blending and starter)

½ tsp (1.4 g) yeast or ¼ cup (60 g) sourdough starter

5 cups (600 g) whole einkorn flour, plus ½ cup (60 g) for starter

½ cup maple syrup (160 g) or honey (170 g)

1 tsp (6 g) salt

¼ cup (32 g) chia seeds

1¾ cups (420 g) warm spring water for dough

1 Tbsp (15 g) olive oil

Directions

DAY 1

Sprouts and Starter: Immerse the einkorn berries in water, and float off any hulls. Soak einkorn berries overnight in 1½ cups spring water.

Activate Starter: If you already have sourdough starter, feed it by adding in ½ cup einkorn flour with ½ cup spring water.

DAY 2

Bread Dough: The next day simmer the swollen berries in a heavy covered pot or pressure cook till soft; without adding additional water. Lightly blend the sprouts with the maple syrup or honey (add additional tablespoons of spring water as needed) so the sprouts and sweetener partially maintain their shape. Mix together all the dry ingredients: einkorn flour, salt, and chia seeds (plus yeast if you use this). Separately mix together the wet ingredients: starter, warm water, oil, and blended sprouts. Mix together the wet and dry ingredients, and fold the dough sides into the center several times. Let dough rest for a half hour or so. Oil or flour your hands. Form loaf. Dust the outside of the loaf well with flour. Place in an oiled, floured bread pan. Score, and cover with a plastic bag. Let slow-rise in the refrigerator overnight.

DAY 3

Bake the next day at 325°F (163°C). Place a pan of water in the bottom of the oven to increase the humidity for a well-caramelized crust. After 50 minutes, turn off the oven and keep the oven door shut. Remove bread from the oven in half an hour, and let cool. Enjoy!

Einkorn Sprout Bread Variation: For a sproutier bread, rinse the einkorn berries for an additional day, then blend with the maple syrup or honey on day three — without simmering. Mix with the dough.

Rye Sourdough Sprout Variation: Substitute half rye berries and half rye flour for einkorn berries and flour for a hearty, satisfying loaf echoing eastern European flavors.

Seaweed Bread Variation: Enjoy the rich, substantial flavor of seaweed in your einkorn sprout bread. Add the lightest varieties, such as nori or kelp, to the dough. First blend or tear the dried seaweed into small pieces. Simmer the seaweed for about 20 minutes till tender. Drain before adding to the dough. It is my husband's favorite bread of all!

Einkorn Sandwich Bread

Light, fluffy, and perfect for sandwiches.

Ingredients

4 cups (480 g) sifted einkorn flour plus more flour to dust work surface
1 tsp (6 g) salt
1 tsp yeast (3 g) or ¼ cup (60 g) sourdough starter with additional
    1 Tbsp (15 g) flour
1 egg (50 g)
2 Tbsp (30 g) olive oil, melted butter, or heavy cream

2 Tbsp honey (42 g) or maple syrup (40 g)

1 Tbsp (15 g) vanilla extract

1 cup (240 g) warm milk

Directions

Mix dry ingredients: flour, salt, and yeast (if you are using it). In a separate bowl, whisk or blend the egg, olive oil, honey, vanilla extract, warm milk, and sourdough starter (if you are using it) together. Mix liquids into dry ingredients. Let the mixture rest for about 15 minutes so the liquids and fat are well absorbed into the einkorn flour. Dust work surface with flour. Knead lightly, let rest for a half hour, then flatten the dough, and fold the edges of the dough to the center for a few times. Let rest for an hour in a cool place. Shape the dough, and coat with flour. Place in an oiled, flour-dusted bread pan. Score the loaves with a sharp knife. Cover with plastic. Let slow-rise in a cool, dark place for 2 or 3 hours for the yeasted recipe or 4 hours for the sourdough. (Better yet, let it rise overnight in the refrigerator.) Place a pan with water in the bottom of the oven to add steam during baking for a crispier crust. Preheat the oven to 450°F (232°C). Remove the plastic. Bake at 350°F (177°C) for about 50 minutes till golden brown. For a crispier crust, let the bread sit in the oven for a half hour after the heat is turned off. Let cool fully before slicing.

⤜⟫⟫⟫⟫⟩ *Saluf: Yemenite Fermented* ⟨⟨⟨⟨⟨⟨⤙ *Pan Bread with Fenugreek and Chickpeas*

Fenugreek seeds, a beloved Yemenite spice known as *hilba*, are a traditional cure for digestive problems and to promote breast milk flow for nursing mothers.

Ingredients

1 heaping Tbsp (15 g) sprouted blended chickpeas

1 heaping Tbsp (15 g) sprouted blended fenugreek seeds

2 cups (240 g) sifted einkorn flour

½ cup (120 g) refreshed sourdough starter

1 cup (240 g) spring water

1 egg, beaten (50 g)

½ tsp (3 g) salt

1 Tbsp oil

½ cup (100 g) grated goat cheese and diced greens for topping, optional

Directions

Soak the chickpeas and fenugreek seeds in water. Rinse the next day. Before chickpea and fenugreek rootlets start to emerge, blend together with the flour, sourdough starter, and water to a pancake batter consistency. Cover the bowl, and place in the refrigerator over-night to ferment. Beat in an egg the next day. Add salt. Heat a frying pan on a medium flame. When the pan is hot, turn down the flame to low, and add oil. Spread dough as thin as possible in the pan. Cover the pan with a lid and pan-bake the saluf on low heat. It will rise thick and bubbly as it bakes. Flip over when golden brown. Grate on goat cheese, and top with sprouts. Served rolled up with a yogurt blended with dill and spices as a dip.

Khachapuri: Cheese Bread

I am enchanted by Georgians, drawn to their warm vibrance like no other culture. This bountiful country has a long, proud history of fierce farmers carrying swords in the fields and radiantly beauti-ful women. The soul of the people is embodied in the invigorating dance, robust polyphonic song, sumptuous foods, and deep love of their land. Georgia is the ancestral homeland for more species of wheat than most of us have heard of, and to the *supra* — bountiful cel-ebratory feasts with noble toasts to God, nature, friends, and heroes, with wine flowing and endless dishes on the table. Khachapuri is a delicious eggy cheese bread enjoyed at the Georgian table.

Ingredients

DOUGH

1½ cups (360 g) yogurt or kefir

½ tsp (3 g) salt

½ tsp (1.4 g) yeast

½ cup (100 g) cheddar cheese, grated

1 Tbsp (21 g) honey, optional

1 beaten egg (50 g)

4 cups (480 g) sifted einkorn flour

1 beaten egg (50 g), for glaze

WILD GREENS-EGG-CHEESE FILLING

1 cup (200 g) grated cheddar cheese

2 beaten eggs (100 g)

2 Tbsp (30 g) yogurt or kefir

1 Tbsp (10 g) finely grated garlic

½ cup (50 g) finely chopped greens such as arugula, nettles, purslane, ramps, or dandelion greens

Directions

To make the dough, mix yogurt, salt, yeast, cheese, and honey if using; whisk in the egg; and knead in the flour to form a soft dough. Divide into two equal parts, and roll each into a large circle. For the filling, mix together the cheese, eggs, yogurt, garlic, and greens.

Spread the cheese filling on top of the dough circles, leaving about an inch from the edges clean. Fold the edges over the top, and pinch firmly together. Shape into a circle. Brush with egg and pierce with a fork. Place the dough circles on a preheated pizza stone (or baking sheet if you do not have one) covered with parchment paper. Bake at 375°F (191°C) till golden brown.

BAGELS AND BUBLIKI

Europe abounds in traditions of diverse bread rings that are boiled, then baked for a chewy texture. The first written mention of the bagel is by the Polish Krakow Jewish Council in 1610 in discussion of foods prepared for the celebration of the birth of a child. The Russian *bubliki* round bread is enriched with eggs and cream, first boiled in spiced milk, then baked to a golden hue.

⤖⤖⤖ *Einkorn Bagels* ⬻⬻⬻

This recipe will make a dozen bagels. I usually top the bagels with sautéed garlic and onions.

Ingredients

7 cups (840 g) sifted einkorn flour

2 cups (480 g) warm water

½ tsp (1.4 g) yeast, or ¼ cup (60 g) sourdough starter

1 tsp (6 g) salt

1 tsp (15 g) oil

2 Tbsp (40 g) maple syrup (plus 1 Tbsp to add to the water that the bagels are boiled in)

1 Tbsp (15 g) diastatic or barley malt, optional

1 tsp (8 g) vital gluten, optional

Topping suggestions: diced garlic and onions, sesame or chia seeds

Directions

Mix all ingredients. Knead well. The dough will be tighter than bread dough. Divide into 12 balls.

Shaping: Roll into thick snakes about ¾ inch (2 cm) diameter and 8 inches (20 cm) long. Pinch ends together to form the bagel shape. Or roll into ball. Press your thumb through the center, and stretch out into the bagel shape. Place on floured parchment paper on a baking sheet. Cover with plastic, and refrigerate overnight.

Next Day: Preheat oven to 450°F (232°C). Boil water with maple syrup, which imparts a rich hue to the baked bagel. Place a test bagel in the water. Turn over after about 15 seconds. It should float. If it does not, let the other bagels rise another half hour or so at room temperature. When the test bagel floats, proceed to boil all the bagels, about 30 seconds on each side. Allow sufficient room in the pot for each bagel. Drip-dry on the baking sheet. Sprinkle on toppings, if any. Bake on a parchment-lined sheet for about 20 minutes till golden.

⇒⟫⟫⟫⟫⟫ *Bubliki Russian Bagel* ⟪⟪⟪⟪⟪⇐

A light, rich, sweet bread ring boiled in spiced milk before baking.

Ingredients

2 eggs (100 g), 1 separated

¼ cup honey (85 g) or maple syrup (80 g)

1 Tbsp (15 g) vanilla extract

¾ cup (180 g) whole milk or cream

2 Tbsp (30 g) butter

1 cup (122 g) tapioca flour

2½ cups (300 g) sifted einkorn flour

½ tsp (1.4 g) yeast

½ tsp (3 g) salt

½ cup (50 g) chia, poppy, or sesame seeds

MILK BATH

4 cups (960 g) whole milk

1 Tbsp (15 g) vanilla extract

Pinch of nutmeg

Directions

Combine 1 large egg, honey, vanilla, and whole milk. Cut butter into the flours. Combine wet ingredients with the flour and butter mixture, yeast, and salt. Knead well. Divide the dough into balls. Press your thumb into the center, and spread out the dough to form a ring. Cover with plastic. Let slow-rise overnight in the refrigerator. The next day preheat the oven to 400°F (204°C). Use a pizza stone if available. Boil whole milk spiced with vanilla and nutmeg. Drop in dough rings. Turn after 30 seconds. They are done in about a minute. Remove with a slotted spoon, and drain. Brush with an egg yolk. Top with chia, poppy, or sesame seeds as preferred. Bake about 25 minutes until golden.

Spiced Ricotta Variation: After the bubliki is boiled, add a tablespoon of lemon juice to the hot milk. Simmer and slowly stir till curds form. Remove with a slotted spoon, and drain in a colander. Enjoy this tasty spiced ricotta smeared on the bubliki.

 *Jerusalem Flatbread*

Ingredients

3½ cups (420 g) sifted einkorn flour
½ tsp (1.4 g) dry yeast or ¼ cup (60 g) starter
2 Tbsp (30 g) olive oil
1 tsp (6 g) salt
1 cup (240 g) warm spring water
Za'atar

Directions

Mix einkorn flour, yeast or starter, olive oil, and salt with just enough water to hold together well. Let rest for 15 minutes to absorb the liquids. Fold and knead into a ball. Put in an oiled bowl, cover, and place in fridge overnight. Next day roll out on a well-floured surface. Sprinkle on the *za'atar* (sesame seeds and thyme). Let rise for an hour. Place on a preheated pizza stone. Bake at 375°F (191°C) till golden.

Options: Brush on topping of olive oil, maple syrup or honey and Worcestershire sauce mixed well together. Sprinkle with thyme. Flour the working surface and roll out as thin as possible for crackers.

Persian Flatbread

For crusty-topped traditional Persian flatbreads, combine a tablespoon of flour with a tablespoon of honey or maple syrup, 1 teaspoon of oil, and ½ cup (120 g) of water. Simmer the flour paste, whisking until thickened. When cool, brush over the flatbread dough. Sprinkle with sesame seeds or za'atar spice. Add Worcestershire sauce to the paste mix for a piquant-sweet flavor.

Savory Bread Sticks

To make cheesy bread stick twists: After flattening the dough, brush with olive oil. Sprinkle on your preferred cheese, Fold the dough over on itself. Slice in thin strips. Twist. Bake on parchment paper at 350°F (177°C) till lightly golden.

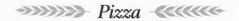 *Pizza*

Ingredients

½ tsp (1.4 g) yeast or ¼ cup (60 g) refreshed sourdough starter

1 cup (240 g) warm water

1 tsp (6 g) salt

3½ cups (420 g) einkorn flour plus more for dusting

1 tsp (5 g) olive oil

1 cup (100 g) grated cheese (to be sprinkled over topping)

Toppings

Directions

If using sourdough starter, refresh the starter a day or two before, and keep refrigerated. Place ¼ cup (60 g) of refreshed starter in a measuring cup. Add warm water until it measures a full cup. Mix all ingredients except toppings together, knead until silky, and form into a ball. Place in an oiled bowl and roll so it is coated with oil. Cover with plastic wrap, and let slow-rise in a cool place for at least 3 hours, or overnight for a fuller flavor. Divide into 2 balls. Roll into flat circles about 14 inches (35 cm) in diameter and a ¼ inch (0.6 cm) thick. Let rise for up to an hour. Add toppings. Bake in the oven at 450° to 500°F (232° to 260°C).

Mezze

Bread, the main food of the traditional Middle Eastern table, was traditionally served with mezze dips and salads throughout the Middle East. The rabbis taught, "One who is about to say the blessings over bread should wait until the salt and mezze (relish) is placed out."[23] Mezze included soups, grain, or vegetable salad drizzled with piquant dressings, olives, onions, cheese, and bean dishes. Meat was typically eaten only a few times a year for festivals or a wedding. Fish was enjoyed in abundance in coastal villages. In times of need or famine, breads were eaten only with salt or vinegar.[24]

New England Spring Risotto

In the chilly New England early spring on my farm, it is thrilling to find fresh vegetables that overwintered under the snow. I made a sublime risotto using the hardy leeks, parsnips, green onions, and parsley that I dug out from the barely thawed soil, with the wild ramps, nettles, lambsquarters, and dandelions gracing the unplowed garden beds, and the onions and garlic stored in our pantry over the long winter.

What is the difference between risotto and pilaf? A moist grain pilaf, slow-cooked in a heavy pot filled with savory meat and vegetables, absorbing the flavors into the soul, holds a central place in tables from Jerusalem to Istanbul, whereas Italian risotto is embraced in a savory broth of intermingled aromas. In 1539, Sultan Suleiman the Magnificent of the Ottoman Empire detailed the foods served at a grand feast that included saffron pilaf, green pilaf colored with spinach and chard juice, red pilaf tinted with grape molasses, and fruit-studded pilaf drizzled with honey and pomegranates. The thirteenth-century Persian Sufi mystic-poet Rumi describes pilafs with rice, chickpeas, mutton, chestnuts, carrots, onions, butter, pine nuts, currants, black pepper, and allspice.[25]

Be it risotto or pilaf, the flavors are deepened when the grains are first sauteed in oil, then simmered in rich chicken or vegetable broth.

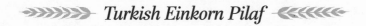

Turkish Einkorn Pilaf

Ingredients

PILAF

2 cups (400 g) einkorn berries

4 cups (940 g) homemade chicken or vegetable broth

2 diced onions (100 g)

4 diced cloves of garlic (25 g)

Fresh leaves of wild spring greens such as wild ramp (wild onion),
     lambsquarters, purslane or stinging nettle—according to availability

1 tsp (6 g) salt

1 Tbsp (15 g) fine white wine

1 Tbsp (15 g) butter or olive oil

DRESSING (OPTIONAL)

½ cup (50 g) diced parsley, dill, and fresh mint

1 cup (240 g) plain kefir or yogurt

Directions

PILAF

Place einkorn in a pot of water. Rinse, and float off the hulls. Soak
einkorn overnight in rich homemade chicken or vegetable broth in
the refrigerator. Saute onion, garlic, and your favorite foraged wild
vegetables in salt, wine, and butter or olive oil. Add einkorn and the
remaining broth. Cover. Slow-simmer in the broth till soft but still al
dente. Spread out the pilaf in a serving dish. Garnish with chopped
scallions or ramps (wild onions) and fresh wild greens as available, or
piquant greens such as arugula.

DRESSING

Blend parsley, dill, and mint in kefir. Drizzle kefir dressing over the
pilaf before serving.

⋙⋙⋙ *Festive Grain Melange* ⋘⋘⋘

A delightful festive dish for a fall harvest festival. Soak einkorn in
water overnight. Simmer in a heavy covered pot. Saute onions, raisins, water chestnuts, salt, and lemon in olive oil. Place on a bed of
cooked einkorn. Garnish with minced cranberries, dates, and apricots softened by simmering in apple cider. Top with roasted chestnuts
or walnuts.

Kreplach and Pasta

Noodles are among the oldest methods to prepare grain for a quick easy food. Around 4,000 years ago a Chinese cook left an overturned bowl of millet noodles that was recently found by archaeologists. The noodles were ⅛ inch (0.3 cm) thick and 20 inches (51 cm) long. Noodles, *itriyot,* are mentioned by rabbis in the Jerusalem Talmud circa the fifth century CE. They discussed whether noodles that were made on the Sabbath could be eaten after the Sabbath. It was decided that noodles could be made on the Sabbath, but only if they are eaten the same day.

In 1154 Moroccan geographer al-Idrisi wrote that Sicilians made a food of flour that they formed into strings that are dried in the sun, then shipped by boat throughout Italy and to other countries. Noodles are served during Nowruz, the Persian New Year, as a symbol of the interwoven web of life and family.

⟫⟫⟫ *Einkorn Kreplach* ⟪⟪⟪

Kreplach are small dumplings filled with a savory filling, often served in chicken soup, that resemble Italian tortellini or Chinese wontons. My aunt Esther from Rajhon, Poland, made the most delicious kreplach I have ever tasted. Her recipe for kreplach dough is perfect; supple and soft enough to roll thin, but dry enough to roll without sticking.

Ingredients
3 cups (360 g) sifted einkorn flour
1 tsp (6 g) salt
4 beaten eggs (200 g)
Filling: 1 cup (100 g) of your choice of savory sauce, meat,
   or grated cheese

Directions
Mix flour and salt in large bowl. Blend or beat eggs well, then mix into flour. Knead into a firm, silky dough. (Since egg sizes and humidity vary, if the dough feels too dry, add a scant teaspoon of warm water as needed. If too moist, add a scant teaspoon or two of flour.) Roll out

very flat on a well-floured surface. Dust the top of the dough with flour. Cut into large squares, or use a cup or cookie cutter to make circles. Fill with savory sauce, meat, or cheese. Personally I prefer cheddar cheese. Fold in half, roll around your finger, then pinch corners together. Let dry for an hour (or freeze when dry). Place in boiling water, and cook until al dente.

Spaghetti Variation: For long, thin spaghetti noodles, after rolling out on a well-floured surface, roll up the flattened, well-floured dough like a rug. Slice thin with a shape knife. Air-dry in cool shade overnight or boil fresh in soup. Einkorn noodles that dry in the shade hold together better than noodles dried in the sun. Makes 5 servings.

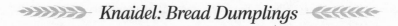

Knaidel: Bread Dumplings

If there is leftover bread, this delicious recipe for knaidel dumplings will give stale bread new life. Keep your hands wet when forming the balls.

Ingredients

1 cup (120 g) einkorn flour (or stale crumbled bread)
Warm water to cover
3 beaten eggs (150 g)
2 Tbsp (30 g) olive oil or melted butter
½ cup (50 g) chopped parsley, scallions, and greens
Pinch of salt
1 scant tsp (3 g) fresh grated ginger to taste
Rich soup broth

Directions

Soak the einkorn flour or stale bread pieces in warm water for 15 minutes until liquid is absorbed. Fold in the beaten eggs, oil, chopped greens, salt, and ginger. Gently shape into walnut-size balls so they just hold together. Simmer in rich soup broth at medium-low heat for at least half an hour.

Nana's Matzah Ball Soup Variation: Soak crumbled einkorn matzah in bubbly seltzer instead of water. Add a tablespoon of olive oil. Add a teaspoon of freshly grated ginger, dash of nutmeg, and diced greens. Serve in rich broth garnished with parsley.

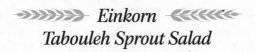

Einkorn Tabouleh Sprout Salad

With arugula, tomato, goat cheese, and a lemon-garlic-olive oil dressing, this is a refreshing dish for a hot summer day.

Ingredients

3 cups (200 g) finely chopped arugula (or spinach), heirloom tomatoes, green scallions, and wild greens such as parsley, purslane, dandelion, and lambsquarters

3 cups (600 g) sprouted einkorn

2 cloves fresh garlic, diced (15 g)

1 Tbsp (10 g) fresh finely chopped thyme

Dash of salt

½ cup (50 g) crumbled feta or goat cheese

Dressing

2 Tbsp (30 g) olive oil

2 Tbsp (30 g) fresh-squeezed lemon juice

1 Tbsp (20 g) maple syrup

2 garlic cloves, finely diced (15 g)

Directions

Chop arugula, tomatoes, scallions, wild greens, garlic, and thyme. Toss with einkorn sprouts. Whisk together the oil, lemon juice, maple syrup, and diced garlic. Drizzle the dressing on. Season with salt. Garnish with the crumbled feta cheese and sprigs of parsley.

How to Sprout Einkorn: Place grains in a bowl of water, swish, and float off the hulls. Soak grains overnight. Pour off water, cover, and rinse twice a day. The sprouts are ready to eat in 48 to 72 hours, just before they sprout rootlets.

 *Hummus*

Hummus is not only delicious but is good for you. My grandmother told me to add raw, finely diced garlic to dishes because it helps chase away colds and infections. She called it "nature's medicine." Garlic contains allicin, an antifungal and antibiotic substance that promotes infection-fighting white blood cells and actually does help reduce infections.

Ingredients

1 cup (200 g) dry chickpeas

6 cups (1440 g) water

1 tsp (7 g) baking soda (enhances creaminess)

3 Tbsp (45 g) tahini

2 Tbsp (30 g) fresh squeezed lemon juice

2 or more cloves crushed raw garlic to taste (15 g)

½ tsp (3 g) salt

3 Tbsp (45 g) olive oil

1 cup (50 g) finely diced parsley or wild-foraged greens for garnish

Directions

Place dry chickpeas in a large bowl, rinse, and cover with water. Soak the chickpeas overnight. Drain. Place chickpeas and baking soda in a heavy pot. Cover with water. Put lid on. Bring to a boil, then simmer about an hour until very soft. Skim off any skins that float to the surface. Blend the cooked chickpeas and chickpea water till smooth. Drizzle in the tahini, lemon juice, raw garlic, and salt. Drip in cold water as needed for a smooth texture. Serve on a plate drizzled with olive oil and diced garlic, and garnish with parsley.

Dulce

An Italian musical term meaning to play gently and sweetly, *dulce* also refers to the subtle flavors of an elegant dessert.

Chocolate

Chocolate is a divine food. We know it, and the Mayans and Aztecs knew it. Age-old traditions for preparing chocolate have been passed down from as long as 5,000 years ago by proto-Mayans who believed that the cacao bean had magical and spiritual powers.

What is the fine art of making chocolate? First, the pulp and seeds of ripened cacao pods are scooped out and spread on mats to ferment for a week in the shade. The resulting liquid is drained away as the temperature rises. The seeds begin to germinate, but high temperature and acidity impede full germination. The seeds are sun-dried for another week, then roasted for about 40 minutes at about 150°F (66°C) to enhance flavor. The timing and temperature of the fermentation is the key to bringing out the finest chocolate flavor. The skin surrounding the seed is removed, leaving the "cocoa nib" of unprocessed, dark, unsweetened chocolate. Ancient chocolate recipes for drinks, porridges, and condiments combine cacao with milk, sugar, nuts, and spices such as vanilla, chili peppers, honey, annatto (a South American spice), and allspice to create exotic delicacies and potent fermented beverages.

The ancient fertility goddess of chocolate, flowers, and fruits, known as Ixcacao, was an earth goddess responsible for blessing the harvest and protecting the people. Legends of cacao are recounted in the Popol Vuh, a sacred book of the K'iche' Maya of Guatemala. It is told that the gods attempted to create humans, but the early people did not survive. When the gods found the right foods to sustain humans — maize and cacao from the Mountain of Sustenance — the people thrived.

Chocolate contains chemicals that include phenylethylamine, an amphetamine-like substance that elevates brain chemicals associated with pleasure; anandamide, which activates receptors that give enhanced feelings of well-being; and theobromine, a mild stimulant similar to caffeine, but with less potency. Their effects include myocardial stimulant, diuretic, smooth muscle relaxant, and coronary artery dilator.[26]

 *Chocolate Babka*

Ingredients

GANACHE FILLING

1 cup (240 g) heavy cream

4 Tbsp (60 g) butter

2 Tbsp (30 g) vanilla extract

¾ cup (238 g) maple syrup

2 cups (200 g) dark chocolate drops

1 Tbsp (15 g) cocoa powder

DOUGH

3 eggs (150 g)

4 Tbsp (60 g) butter

1 Tbsp (15 g) vanilla

1 cup (240 g) heavy cream

1 cup (317 g) maple syrup

1 Tbsp (15 g) olive oil

5½ cups (660 g) sifted einkorn flour

½ cup (61 g) tapioca flour

½ tsp (3 g) salt

½ tsp (1.4 g) yeast

STREUSEL TOPPING

½ cup (60 g) sifted einkorn flour

2 Tbsp (30 g) butter

3 Tbsp maple syrup (60 g) or honey (63 g)

½ tsp (3 g) salt

Directions

GANACHE FILLING

Melt ingredients on low heat, stirring frequently. Spread onto the flattened dough while slightly warm.

DOUGH

Mix dry ingredients together. Put all wet ingredients in a blender. Pulse until mixed. Mix into dry ingredients. Let rest 15 minutes, then fold and knead. Let slow-rise in a cool place for an hour. Divide dough into 5 balls. Dust working surface generously with flour. Flatten each ball into round circle about ¼ inch (0.6 cm) thick. Slather with the ganache filling. Roll up like a rug. Twist into an interesting shape, then form the twisted dough into a loaf shape. Place in a small oiled baking pan that is dusted with flour. Slice open the top rolls artistically to expose the inner chocolate. Brush with beaten egg or garnish with streusel. Place in a plastic bag, and let low-rise overnight in the refrigerator. The next day, bake at 325°F (163°C) for about half an hour until golden. Optional: Brush with maple syrup immediately on removing from the oven for a crusty sweet glaze.

STREUSEL TOPPING

Mix the flour, butter, sweetener, and salt together into crumbs. Spread on the top of the rolled chocolate-filled dough.

 *Shiker Babka*

Shiker means "drunk" in Yiddish and "liquor" in biblical Hebrew.

Ingredients

1 babka hot from the oven
1 cup maple syrup (317 g) or honey (340 g)
1 cup (240 ml) of fine rum, brandy, whiskey, or your favorite spirits

Directions

As the babka is baking, boil and whisk the sweetener and spirits together to make a thick syrup. As soon as you remove the babka from the oven, poke holes in it with a fork and slowly spoon the liquid on it, taking care that each spoonful is absorbed before the next is added, till all is absorbed, making a moist cake drenched with the syrup.

Chocolate Biscotti (or Mandelbrodt)

Biscotti, known in Yiddish as mandelbrodt, are twice-baked, crunchy biscuity-bready scones enjoyed with coffee or tea, or packed for a journey. My grandmother used to bring a shoebox brimming with sublime homemade mandelbrodt wrapped in brown paper when she came to visit. While mandelbrodt and biscotti both have a crunchy crust, mandelbrodt is slightly softer than biscotti due to its more generous butter content. Mandelbrodt was popular in eastern Europe among traveling storytellers and itinerant merchants as a sweet bread that kept well on the road. Baking mandelbrodt or biscotti is an auspicious time for storytelling and folklore with young people.[27] In earlier times, twice-baked breads were a staple food in times of war or for long journeys.

Ingredients

1 cup maple sugar (144 g), honey (340 g), or organic sugar (200 g)

8 Tbsp (120 g or 1 stick) melted butter or cream

1 Tbsp (15 g) brandy

4 Tbsp (60 g) vanilla extract

4 eggs (200 g)

4 cups (480 g) sifted einkorn flour

1 heaping tsp (6 g) baking powder

¼ tsp (1.5 g) salt

¼ cup (32 g) chia seeds

1 cup (200 g) chocolate chips

Dash nutmeg, cinnamon, and/or ginger

1 cup (200 g) walnuts and dried fruit

Directions

Blend sweetener, butter, brandy, vanilla, and eggs together. Mix dry ingredients: flour, baking powder, salt, chia seeds, chocolate chips, spices, nuts, and dried fruit. Fold into wet ingredients. The thick batter will be too wet to shape with your hands. Pour into oiled and floured small bread pans to just half full. If you do not have bread pans, add more flour to make a wet dough and form a flat, oval loaf on a baking sheet covered with parchment paper. Place in the oven and set to 350°F (177°C; do not preheat). Bake until lightly golden, about 45 minutes. Turn off oven. Let sit in the warm oven another half hour. Remove loaves from oven. Let cool well. Cut loaves into ½-inch slices with a wet serrated knife. Arrange slices on the baking rack in the oven with a baking sheet underneath to catch any crumbs. Bake at 225°F (107°C) for an hour or so till dry and golden. Let cool completely. Store in an airtight container at room temperature.

Variations: Substitute 1 cup (122 g) tapioca flour for 1 cup (120 g) einkorn flour for a lighter crispness. Add a tablespoon of brandy and a tablespoon of einkorn flour. The eastern European Jewish version includes generous almonds, walnuts, raisins, and cinnamon.

Sculpty Dough for Pastry Art

Ingredients
4½ cups (540 g) sifted einkorn flour
Pinch of baking powder
Pinch of salt
8 Tbsp (120 g or 1 stick) butter
1 cup maple syrup (317 g) or honey (340 g)
4 Tbsp (60 g) vanilla extract
1 egg (50 g)

Directions
Mix dry ingredients together. Blend butter, maple syrup, vanilla, and egg together. Mix with the dry ingredients. Knead. Chill in the refrigerator for an hour. Use as a sculpty medium to shape. I rolled the dough flat, cut circles, placed a dollop of chocolate ganache on each, then folded and shaped. Bake at 325°F (163°C) till lightly golden.

⫸⫸⫸ *Pastiera Ricotta Cheesecake* ⫷⫷⫷

Pastiera is a festive spring cheesecake made with fresh-harvested wheat, eggs, and cheese baked since early pagan times, enjoyed by Italian Jews for the Shavuot spring wheat festival and in Easter spring celebrations to this day.[28]

The pastiera cake of sprouted wheat, eggs, and ricotta cheese evolved from a pagan celebration of spring abundance when priestesses of Ceres carried eggs, freshly harvested wheat, and milky dishes in a festive procession from spring birthing as symbols of renewal. It is also told that the sea spirit Parthenope would emerge from the waters of the Gulf of Naples every spring to sing enchanting songs. Deeply moved by the haunting beauty of the melodies, the people gathered at the shore to listen. In gratitude for the sweetness of the songs, the people offered Parthenope the most cherished products of their land. Beautiful maidens brought gifts of fresh-harvested wheat, which expressed the strength and richness of the land; ricotta cheese, a gift from the shepherds and animals; eggs, a symbol of renewed life; flower blossom water mixed with sweet spices, and honey, like the sweetness of Parthenope's voice.

Parthenope, pleased with the gifts, brought them to the gods and goddesses, who combined everything together with a heavenly art that embodied the sweetness of the season, creating the first

Cucina Hebraica

There is a small, very old shop in the Roman ghetto that sells einkorn cheese *crostata* in pieces, along with *castagnaccio* made with chestnut flour, raisins, pine nuts, and fennel seeds, and a pizza made with polenta, pine nuts, raisins, and sugar, plus other traditional Roman pastries. The Roman Jews consider themselves the most authentic descendants of antique Roman cooking, and to a great extent it is true. Their recipes seem to have remained pure and closer to the traditional methods.[29]

206

Pastiera Napoletana. Another legend tells that fishermen's wives would go to the shore to leave gift baskets with ricotta, wheat, eggs, and flowers as offerings for the sea spirits in hopes that their husbands would come back safely. In the morning when the wives returned they saw that during the night the ingredients had been mysteriously mixed together into a pastiera cake in their baskets. In ancient Rome, an einkorn-ricotta pastiera was eaten at the *confarreatio*, a symbol of fertility for the ancient Roman wedding ceremony. In the old recipes, sprouted einkorn was lightly boiled to delicate tenderness.

Ingredients

EINKORN PIECRUST

3 cups and 1 heaping Tbsp (375 g) sifted einkorn flour

3 sticks (360 g) grated unsalted butter

Juice of 1 lemon (45 g)

½ tsp (3 g) salt

1 egg yolk (25 g)

¼ cup (60 g) heavy cream

1 Tbsp (15 g) vanilla extract

¼ cup (80 g) maple syrup

Optional: Finely crushed pine nuts or walnuts

FILLING

1 cup (200 g) sprouted einkorn (see page 164)

2 cups (400 g) fresh homemade ricotta

4 beaten eggs (200 g)

¼ cup (60 g) heavy cream

⅓ cup maple syrup (106 g) or honey (113 g)

¼ cup (59 ml) of liqueur such as brandy, marsala, or rum, or homemade fruit or herb spirits

Scant ¼ tsp (1.5 g) salt

1 Tbsp (10 g) lemon zest

1 Tbsp (15 g) vanilla extract

Juice of 1 lemon (45 g)

Pinch of nutmeg

Directions

CRUST

The key for light, flaky piecrust is to use frigidly cold ingredients. Chill the flour, butter, juice, and bowl in the refrigerator an hour before mixing. Mix together the flour and salt. Grate in the sticks of cold butter, and mix into the flour by hand. Mix the egg yolks, lemon juice, cream, vanilla, and maple syrup together. Add the liquids to the flour mixture a tablespoon at a time until it holds together. The exact amount is determined by the humidity and the flour. Do not overwork it. Form the dough into a disk and wrap in plastic. Refrigerate for an hour so the moisture is well absorbed. Dust the working surface and rolling pin with flour. Roll out the dough on well-floured parchment paper, then transfer to pie plate. If you do not have parchment paper, roll the

Homemade Ricotta: Where There's a Will, There's a Whey!

Making fresh ricotta is as easy as brewing a pot of herb tea. Once you see how simple it is, the insipid commercial ricottas will be a bland, faded memory.

Use the freshest whole organic milk you can find. Stir in a tablespoon of fresh-squeezed lemon juice per half gallon (2 L) of milk. Add ½ tsp (3 g) salt per gallon of milk. Heat milk slowly in a heavy pot on low-medium, stirring the bottom to prevent scorching When the milk reaches 165° to 170°F (74° to 77°C), small flakes should start to form and separate into curds. If you do not see the flakes forming, add lemon juice one teaspoon at a time until the curds form. Stir slowly from the bottom to avoid breaking curds that have formed. Over-stirring will cause smaller curds to form. Heat to 190°F (88°C), then turn the heat off. As the curds rise, move them from the sides to the center of the pot with a perforated spoon. Let curds rest for 10 to 15 minutes in liquid whey to enrich the final quality. Ladle curds into a colander to drain. For a lighter ricotta, drain briefly, then chill. For a dense texture, drain several hours before refrigeration.

dough up onto the pin. Lay the dough snugly into a springform or pie pan. Flute the excess dough around the rim. Freeze till ready to use.

FILLING

Float off any hulls. Soak the einkorn for a day in spring water. Simmer the swollen einkorn in milk till tender but not broken. Combine the cool einkorn with homemade ricotta cheese, 4 beaten eggs, heavy cream, maple syrup, liqueur, salt, lemon zest, vanilla, and lemon juice. Add the nutmeg to taste. Fill the pie crust. Crisscross thin strips of pie dough on the top for a decorative lattice. Bake at 350°F (177°C) till golden. Lightly drizzle flour atop the baked pie for a decorative effect.

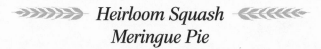

Heirloom Squash Meringue Pie

This luscious and nutritious pie is perfect for a harvest celebration and all year-round. It looks fancy but is so easy to make.

CR Lawn says that the sweetest heirloom squashes are Buttercup, Candy Roaster, an old Appalachian variety (not the North Georgia Candy Roaster that looks like a banana), Sibley, Lower Salmon River, Sunshine, and Eastern Rise. I love the Buttercup best.

Ingredients

PIECRUST

1 egg (50 g)
3 Tbsp (45 g) icy juice (or 2 Tbsp [40 g] maple syrup and
   1 Tbsp [15 g] vanilla)
1 stick (120 g) cold unsalted butter
1½ cups (180 g) chilled sifted einkorn flour
Scant ½ tsp (3 g) salt
Option: substitute ½ cup (61 g) tapioca flour for ½ (60 g) cup einkorn flour

FILLING

Buttercup squash
3 eggs (150 g)
1 cup (240 g) heavy cream

1 Tbsp (15 g) vanilla extract

Pinch sea salt

Dash of ginger, cinnamon, nutmeg to taste

½ cup (100 g) pecans or chestnuts

MAPLE-NUTMEG MERINGUE

5 eggs (250 g), separated

½ tsp (3 g) cream of tartar

½ cup (160 g) maple syrup

1 Tbsp (15 g) lemon juice, optional

1 Tbsp (15 g) vanilla extract, optional

Light sprinkle of nutmeg

Directions

PIECRUST

Whisk egg with liquids. Grate in cold butter. Work in the flour and salt. Press the dough into a disk. Wrap in plastic wrap, and refrigerate for at least 30 minutes or up to 24 hours. Preheat the oven to 425°F (218°C). Roll out the dough onto floured parchment paper. Turn into floured pie plate. Prick bottom with a fork. Trim and crimp the edges. Refrigerate for 30 minutes. Line the chilled dough-lined pie pan with aluminum foil with generous overhang. Pour dried beans or pie weights on the aluminum foil up to the top of the crust. Bake the piecrust until the dough has set for about 25 minutes.

FILLING

Cover a whole squash in foil, and bake it in a large pan with an inch of water at 350°F (177°C). Take out of the oven when it is soft – about an hour. When it has cooled, scoop out the inside of the baked squash. Blend with the eggs, heavy cream, vanilla, salt, and spices. Place the filling in the piecrust, and top with nuts.

MAPLE-NUTMEG MERINGUE

The mystic of meringue may be daunting, but it is an easy, fun, and nutritious addition to any dessert. Let eggs warm to room temperature. Separate egg whites, making sure no yolk falls in. Whisk egg whites with cream of tartar until peaks form. Cream of tartar stabilizes the meringue. This is one of the few times I use an electric mixer. After the froth is well formed, slowly whisk in maple syrup and lemon juice or vanilla (optional) until the meringue gets stiff shiny peaks. Lightly prebake the piecrust with the squash filling, then spread the meringue on the hot filling, making delightful peaks on top. Or put the meringue in a plastic bag with a corner cut off and squeeze out delicate shapes. Bake at 325°F (163°C) for about 20 minutes till the peaks are lightly browned.

To make meringue cookies, place dollops on parchment paper. Bake at 225°F (107°C) for an hour and a half, then let sit in the oven cooling for a half hour.

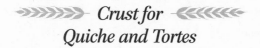

Crust for Quiche and Tortes

Ingredients

1 beaten egg (50 g)

1 Tbsp maple syrup (20 g) or honey (21 g)

1 Tbsp (15 g) vanilla extract

1 Tbsp (15 g) lemon or orange juice

1 Tbsp (15 g) heavy cream

8 Tbsp (1 stick or 120 g) cold salted butter

2 cups (240g) plus 1 Tbsp (15 g) sifted einkorn flour with extra
    to dust working surface

¼ tsp (1.5 g) salt

Directions

Blend together the egg, maple syrup, vanilla, juice, and cream. Grate the cold stick of butter into the flour. Add salt. Mix in the blended liquids. Roll out on generously floured parchment paper. Place in freezer for half an hour if too soft. Hold parchment paper upside down over the pie dish and peel off the piecrust dough into the dish. Trim and flute edges. Freeze for an hour. Cut a few slashes in the bottom of the piecrust to let air out as it bakes. Place a sheet of parchment paper over dough in the pie plate and fill with dried beans. Preheat oven to 375°F (191°C), and bake 15 to 20 minutes, until firm. Remove parchment and weights. Let cool. Add filling, and bake according to filling recipe.

Variation: Substitute ½ cup (61 g) tapioca flour for ½ cup (60 g) einkorn flour for enhanced crispness.

Flakier Crust Variation: 2 cups (240 g) sifted einkorn flour, 16 Tbsp (2 sticks or 240 g) cold butter grated into flour, pinch salt, 3 Tbsp (45 g) icy-cold orange or lemon juice.

Light, Lower-Butter Crust Variation: 2 cups (240 g) sifted einkorn flour, 8 Tbsp (1 stick or 120 g) cold butter grated into the flour, 2 beaten eggs (100 g), 1 Tbsp (20 g) maple syrup, 1 Tbsp (15 g) vanilla extract, ½ tsp (3 g) salt.

Maple Vanilla Einkorn Cookies

Ingredients

½ cup (120 g) heavy cream or 8 Tbsp (1 stick or 120 g)
    chilled butter, grated

3 cups (360 g) sifted einkorn flour

1 tsp (6 g) sea salt

1 tsp (5 g) baking powder

½ tsp (1.4 g) yeast

½ cup (60 g) chia seed

1 whole egg and 1 egg white (75 g)

½ cup (160 g) maple syrup

2 Tbsp (30 g) vanilla extract

Directions

If using grated cold butter, work into the flour. Combine the salt, baking powder, yeast, chia seeds, and the flour-butter mixture. Whisk together the eggs, syrup, and vanilla (and cream if you are using this). Mix together the wet and dry ingredients. Let rest half an hour. Oil a cookie sheet and dust with flour, or place on parchment paper. Drop cookie batter onto sheet. Bake for about 15 minutes at 350°F (177°C) till golden.

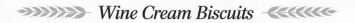

Wine Cream Biscuits

A light, crispy biscuit with subtle creamy richness infused with a delicate wine aroma that can be enjoyed with coffee or at a festive wine and cheese table.

Ingredients

½ cup maple sugar (72 g) or organic sugar (100 g)
1½ cups (180 g) sifted einkorn flour
½ cup (61 g) tapioca flour
½ tsp (3 g) salt
Pinch baking powder
½ cup (120 g) heavy cream
¼ cup (60 g) plus 1 Tbsp (15 g) white wine
1 Tbsp (15 g) maple syrup or honey, for glaze
1 Tbsp (15 g) olive oil or melted butter, for glaze

Directions

Combine all ingredients. Knead well. Shape. Glaze with a mix of maple syrup or honey and olive oil or melted butter. Let rest for half an hour. Bake in a preheated oven set to 350°F (177°C). Bake for about 25 minutes till golden. Lightly dust with flour.

⇶⇶⇶ *Savory* ⇷⇷⇷
Garlic Cheddar Popovers

Popovers' crispy outside and fluffy light inside are an easy, elegant dish. They can be whipped up quickly for a nutritious breakfast treat or served as sumptuous dinner rolls. This recipe makes six popovers.

Ingredients

Olive oil to brush wells

1 cup (120 g) sifted einkorn flour

1 cup (240 mL or 240 g) whole milk

2 eggs (100 g)

3 cloves minced garlic (25 g)

½ tsp (3 g) salt

½ tsp (2 g) baking powder

1 cup (120 g) grated cheddar cheese

3 Tbsp (45 g) diced scallions, parsley, spinach

1 Tbsp (15 g) olive oil or melted butter

Directions

Preheat oven to 450°F (232°C). Brush wells with oil. Dust with flour. Whisk together the milk, eggs, minced garlic, and salt till frothy. Slowly whisk in flour, baking powder, ½ cup cheese, 1 tablespoon greens, and melted butter. Let rest at room temperature for an hour. Preheat pan for 5 minutes before filling. This helps popovers rise better. Pour batter into wells ⅔ full. Sprinkle remaining grated cheese and greens on top. Bake at 450°F (232°C) for 10 minutes, then turn down the oven to 350°F (177°C), and bake for 20 minutes until golden. Do not open the oven till they are done or the steam will escape and the popovers may collapse. Remove from the oven, and poke a tiny hole on the side of each popover to release steam so that they do not collapse. Take out of pan immediately. Serve hot.

Variation: Add ¼ cup (30.5 g) tapioca flour for ¼ cup (30 g) of einkorn flour for a crispier texture. Add a tablespoon (15 g) of chia seeds to enhance moistness.

Sweet Popovers Variation: Combine the flour, milk, eggs, salt, baking powder, and butter with a tablespoon of honey or maple syrup.

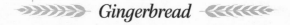 *Gingerbread*

Ingredients

2¼ cups (270 g) einkorn flour

½ tsp (3 g) salt

1 tsp (4 g) ginger powder

1 tsp (4 g) nutmeg

½ tsp (3 g) baking soda

1 tsp (5 g) baking powder

½ cup (100 g) raisins

½ cup (100 g) dried dates (soak in the vanilla extract to soften)

2 Tbsp (30 g) vanilla extract

½ cup (100 g) chopped walnuts

1 Tbsp (15 g) chia seeds

¼ cup (60 g or 4 Tbsp or ½ stick) cold butter

½ cup (112 g) plain kefir or yogurt

2 eggs (100 g)

½ cup (160 g) maple syrup

FROSTING

1 cup (200 g) chèvre soft goat cheese (or cream cheese)

½ cup (160 g) maple syrup

Directions

Preheat oven to 350°F (177°C). Mix all the dry ingredients together: flour, salt, spices, baking soda, baking powder, fruit, nuts, and chia seeds. Grate cold butter into the flour mix. Mix in yogurt, vanilla, eggs, and maple syrup. Place in a small bread pan for gingerbread and in a square pan for cake. Chill the dough for 30 minutes in the refrigerator. Bake at 350°F (177°C) for about 40 minutes. Mix together the goat chèvre and maple syrup. When gingerbread is cool, slather with the maple-cheese frosting.

Kichel

Airy and crunchy, a *kichel* (Yiddish for cookie or biscuit; the plural is *kichlach*) is a light cookie or cracker beloved in eastern Europe and contemporary Israel and becoming popular in the United States. They are delightful for breakfast and for a snack anytime.

Ingredients

4 eggs (200 g)

1 Tbsp (15 g) olive oil, and more for brushing

1 Tbsp (15 g) vanilla extract

3 Tbsp honey (63 g) or maple syrup (60 g), and more for brushing

1 cup plus 2 Tbsp (150 g) einkorn flour

½ tsp (3 g) salt

½ tsp (3 g) baking soda

½ tsp (2 g) baking powder

Directions

Vigorously whisk or blend together the eggs, oil, vanilla, and sweetener. Mix together the flour, salt, and baking soda and baking powder. Mix wet and dry ingredients together. Let ingredients rest for 15 minutes to absorb the liquids and fats. Fold and knead lightly. Let rest for half an hour. Preheat oven to 450°F (232°C). Roll out on a floured surface, and cut into strips 1 inch (2.5 cm) wide and 3 inches (7.5 cm) long. Place parchment paper on a baking sheet. Brush with maple syrup or honey. Lay out the strips on the baking sheet, giving each a twist. Reduce heat to 400°F (204°C). Bake for 5 minutes. Reduce the heat to 300°F (149°C) for about 6 more minutes, until lightly golden. Sprinkle with flour after removing from the oven.

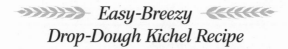

Easy-Breezy
Drop-Dough Kichel Recipe

Ingredients

2 cups (240 g) einkorn flour

6 eggs (300 g)

8 Tbsp (1 stick or 120 g) cold butter, grated

½ cup honey (170 g) or maple syrup (160 g)

½ tsp (3 g) salt

1 Tbsp (15 g) vanilla extract

Directions

Simply mix all ingredients together, let rest for half an hour, and drop the wet dough onto floured parchment paper on a baking sheet. Bake at 350°F (177°C) till golden.

CR Lawn's
Einkorn Ricotta Blintzes

CR recounts, "My grandma Sarah from New Jersey introduced me to the joy of blintzes when I was fourteen. They are my favorite family heritage dish. I modified the recipe over the years, replacing sugar with maple syrup, and lemon rind with freshly squeezed lemon juice. The original recipe called for cottage cheese, but I found ricotta to be vastly superior. Recently I discovered that blintzes are even better made with fresh homemade ricotta. After more than 40 years using unbleached white or whole wheat pastry flour, it was a revelation to discover that einkorn flour has the absolutely perfect texture for making the blintzes so that they really are delicate, luscious crepes that hold together well. You can use jam, lightly cooked blueberries, or apples for the filling, but I prefer lemon juice mixed with maple syrup in ricotta." This recipe makes about 20 crepes.

Ingredients

BATTER

1½ cups (180 g) sifted einkorn flour

¼ tsp (1.5 g) salt

1 cup (240 g) water

3 eggs (150 g)

⅔ cup (161 g) whole milk

4 to 6 Tbsp (60 to 90 g) butter to refresh the pan as needed

FILLING

1 pound (454 g) homemade ricotta

1 egg (50 g)

2 Tbsp maple syrup (20 g) or honey (21 g)

Juice of one small or half a large lemon—adjust to taste

Dash of cinnamon or nutmeg to taste

Directions

BATTER

Mix flour and salt. Add water. Mix to a smooth paste. Add eggs and beat. Blend in milk. Rub the bottom of a skillet with butter. Heat the pan, and add another teaspoon of butter. When the skillet is hot (if batter sticks to the skillet, it is not hot enough), pour 2 to 3 tablespoons of batter into the skillet, just enough to lightly coat the pan for a delicate crepe, not a pancake. Rotate the pan to spread out the batter thinly. Cook until lightly golden on one side. This should take only seconds. Remove pan from the heat, and flip-slide the crepe onto a plate. The crepe should come off cleanly. Refresh the pan with butter and continue.

FILLING

Mix ingredients. Adjust to taste. The filling should be a moist paste, not stiff or runny. Place a heaping tablespoon of filling in the center of the cooked side of the crepe, fold over from both sides, then fold over the remaining ends to get a 3-inch (7.5 cm) square. Fry on both sides till both sides are crisp and golden, or bake in a 400°F (204°C) oven for about 15 minutes till lightly brown. Crepes will rise in the oven but fall again after cooling. I serve them drizzled with maple syrup, but there are many other options.

Acknowledgments

M y years of exploration and discovery that created this book were brought to life by a team effort of peasant farmers, researchers, family, and friends. I want to especially thank CR Lawn, contributing editor and coconspirator; Shulamit Falik, trusty collaborator and baking partner; Vivian Brill for her incredible friendship and delicious food during all my years in the Middle Eastern fields; and my children, Noah and Ezra, for their humor and support in every aspect!

Klaas Martens, coauthor of *Landrace Grain Husbandry*, contributed his vast experience, spiritual inspiration, and love of the soil and the soul for significant parts of chapter 3. Mark Fulford, the soil doctor, contributed his great understanding of holistic soil management and the whole plant. Mariam Jorjadze and Elene Shatberashvili of Elkana Georgian Biological Farming Association (elkana.org.ge) generously contributed deep friendship, as well as dedication to restoring the threatened biodiversity of the land of Georgia and her peasant farmers and generously shared the research of Dr. Taiul Berishvili. Dr. Karl-Josef Mueller's (darzau.de) clarity, experience, and contribution of his breeding lines illumined my understanding of holistic wheat breeding. Great gratitude to Anders Borgen's (agrologica .dk) generous sharing of landrace seed and knowledge, Dr. Kostas Koutis (aegilops.gr) for his grand generosity, Jean Francois Berthelot, Nicolas Supiot, Dr. Dominique Declaux, and Dominique Guillet of Kokopelli — bless his tenacity, love of biodiversity, and the farmers! Dr. Harold Bockelman's devotion to the service of cereal biodiversity, Dr. Abdullah Jaradat's mentorship, Salvatore Ceccarelli's practical inspiration, Dr. Dean Spaner's humor, Glenn and Kay Roberts' support and inspiration to restore landraces (ansonmills.com), Daniella Malin (greenmountainflour.com), her beautiful family and small-scale mill in Vermont — it can be done! Sharon Rempel, Jennifer Greenberg, Heather Darby, and Jack Lazor; John Melquist, baker extraordinaire; Ben Gleason's generosity; Rachel and Tevis Robertson-Goldberg (crabapplefarm.org); Masoud Hashemi — light of UMass; Andrea Stanley (valleymalt.com); Dusty Dowse and Amber

219

Lambke (kneadingconference.com); Jim Amaral — Borealis Breads, Sara and Matt Williams; Aurora Mills; Heather Spalding — the hardworking heart of MOFGA; Sandy and Michael Jubinsky (stoneturtlebaking.com); June Russell (grownyc.org), who works tirelessly to bring local flour to the people; Ben Watson, who grasped the value of landrace wheat from the start; and the Chelsea Green team to bring this work to you.

Now it is up to each of us "to eat it to save it," to gather the seeds for our community, and to grow local seed friends. It takes a community to grow a loaf of landrace bread.

Baker's Formulas

Professional bakers use formulas with proportions so that they can scale up or down or substitute ingredients to create new recipes with consistent success. In the baker's formula, all of the ingredients are in a relative ratio to the weight of the flour, which is always 100 percent. For example, in an artisan bread:

Flour: 100%
Water: 70%
Salt: 2%
Instant yeast: 0.5%
Total = 172.5%

Starting with a base of 1,000 grams of flour, the weight of the other ingredients would be as follows:

Water: 1,000/ 70 = 700 grams
Salt: 1,000/.02 = 20 grams
Instant yeast: 1,000/.05 = 5 grams

Basic Einkorn Bread Formula

Ingredients	Bakers' %	Grams	Schedule
Total Flour	100%	1,000	
Poolish			
Einkorn Flour	45%	450	Day 1 afternoon.
Water	45%	450	Refrigerate overnight.
Yeast	0.05%	0.5	
			Day 2. Refresh poolish in morning.
Dough			
Poolish	All	900.5	Day 2. An hour later,
Einkorn Flour	55%	550	mix everything together.
Water	25%	250	
Yeast (optional)	0.95%	9.5	
Salt	2%	20	
			Let ferment in a cool place for at least 2 hours.
Total Dough (g)		1,730	Form loaves. Let rest half an hour.
Loaves		2 large or 3 medium loaves	Bake.

Dusty Dowse's Einkorn Pita Formula

	Bakers' %	Grams
Total Flour Base		1,000
Levain		
Active Levain	20%	50 (25 water and 25 flour)
Einkorn Flour	100%	250
Water	100%	250
Total Weight		550
Dough		
Levain	All	550
Einkorn Flour	100%	750
Water	58%	435
Olive Oil	5%	50
Salt	2%	20
Total Dough (g)		1805
Total Hydration	69%	
Number Pitas		15

DUSTY DOWSE'S EINKORN PITA

Dusty is the baking educator for the renowned Kneading Conference of Maine (kneadingconference.com), which sponsors an inspiring annual bread conference.

THE SCIENCE OF SOURDOUGH

For those who seek a scientific approach to sourdough bread baking, Raymond Calvel, a renowned French baker, has refined the process to a high art in *The Taste of Bread:*

> We should always be aware that the method of preparing sourdough leavening has an important relationship to taste of the bread. Taste is influenced to an even greater degree by the rate of intensification of mixing and consequent over-oxidation of the dough, and it may be greatly damaged by these factors.[1]

Raymond Calvel's sourdough starter process takes about 2½ days. He incubates the starter at 81°F (27°C), uses salt and malt in the starter, and stores it in the fridge at about 50°F (10°C) to preserve the flavor. Calvel reports that storing sourdough starter below 46°F (8°C) damages the fullness of the natural aroma, even though it maintains its fermentation and ability to rise.

Ingredients

flour (einkorn or rye is good)
dried malt extract (from a brewing house or health-food store)
salt
water
Note: A cup of flour weighs about 120 grams. A cup of starter weighs
   about 240 grams.

Directions

The organisms we are feeding are already in the flour, so there is no need to expose the starter to the air. Cover the starter between feedings with a cloth. If the starter is uncovered, it will dry out. Maintain the starter by feeding equal amounts of flour and water by weight twice a day if the starter is at room temperature or once a day at 50°F (10°C).

Time	Starter	Flour	Water	Salt	Malt
Start	0 grams	300 grams rye and 300 grams wheat flour	360 grams	3 grams	3 grams
22 hours	300 grams	300 grams flour	180 grams	1 gram	2 grams
7 hours	300 grams	300 grams flour	180 grams	1 gram	
7 hours	300 grams	300 grams flour	180 grams	1 gram	
6 hours	300 grams	300 grams flour	180 grams	1 gram	
6 hours	300 grams	300 grams flour	180 grams	1 gram	

After three days of feeding twice a day, the starter should double its size between feedings and be mature enough to use. In each step after the first, 300 grams of starter are used from the previous stage. For the first two days, half of the starter is discarded since it is not yet mature. When the starter is mature it is good to save all of it.

Amylolytic process refers to the conversion of starch into soluble sugars through enzymes. It is used to brew alcohol from grains. Since grains contain starches but no simple sugars, the sugar needed is produced from starch via the amylolytic enzymes. To brew beer, this is done through malting. Malt is added to increase the amylolytic power of the flour. Salt is added to protect the dough against the action of proteolytic breakdown of protein that may weaken the gluten during the early stages of dough fermentation. This otherwise may damage the dough by softening it excessively, since this first fermentation stage may last for more than 20 hours. For this same reason, salt also fulfills an important role in the renewal or feeding of cultures that are ultimately to be used in the building of a naturally fermented sponge or levain.[2]

The key factors are learning how to read the activity level, that is, bubble formation, and the culture's ability to rise and when to feed it. Feeding too late or too early, or at overly cool temperatures, will decrease the richness of flavor.

APPENDIX 2

Why Modern Wheat
Is Making Us Sick

The end of the human race will be that it will eventually die of civilization.
— RALPH WALDO EMERSON

I am allergic to modern wheat. I suffered miserably from bloating, malabsorption, and indigestion for many years. No doctor could help me. Yet when I removed wheat from my diet, the symptoms vanished. My vitality returned with the added bonus of pounds shed. When I eat einkorn, I have no adverse reactions. Thus I have embarked on the journey to understand why. I am deeply disappointed in the wheat industry for ignoring — even refuting — the reports of the rise in gluten allergies.

The rise in gluten allergies is real. Gluten allergies are increasing at epidemic proportions. In children the rate has skyrocketed fourfold in recent decades.[1] The National Institutes of Health convened its first conference on celiac disease only in 2004, concluding that the condition is "widely unrecognized" and "greatly under diagnosed."[2] In other words, most individuals who suffer allergies to gluten are not diagnosed and suffer needlessly with symptoms ranging from poor nutrient absorption and discomfort to unexplained weight gain and bloating. Sensitivity to gluten affects from 1 to 6 percent of the population today; however, most gluten allergies are not diagnosed. A recent study compared frozen serum samples collected between 1948 and 1952 to serum samples collected at the time of analysis.[3] After verifying the stability of IgA antibodies in the old samples, the

225

investigators concluded that gluten allergies and celiac disease are at least four times more prevalent today than they were sixty years ago.

Celiac disease is a gastrointestinal malabsorptive disorder resulting from the ingestion of gluten. Allergic responses to gluten are highly variable and can include inflammation of the intestinal mucosa, which may result in atrophy of intestinal villi; malabsorption of nutrients; and a variety of clinical manifestations that can include diarrhea, abdominal cramping, pain, and distension. Untreated celiac disease and gluten allergies may lead to vitamin and mineral deficiencies, osteoporosis, and other systemic problems.

Celiac disease and a gluten allergy are two separate responses. The ingestion of gluten may lead to an allergy or celiac disease in susceptible individuals. Celiac disease is gluten intolerance that causes inflammation of the small intestinal mucosa. There are a range of gluten allergies and intolerances. Allergies occur through a pathway in which IgE antibodies cause different symptoms compared with intolerance. Different epitopes — the part of an antigen recognized by the immune system antibodies, B cells or T cells — are involved in allergy and celiac disease. Gluten sensitivity has been discovered only recently, and the pathway is still not well understood. The only treatment for celiac disease is a gluten-free diet.[4]

For years, Eric Lubrano[5] wondered what he was suffering from.

I went to dermatologists for my skin problems. They offered me one skin cream, then another. I went to doctors for my stomach pains and poor digestion. Nothing helped. Finally after years of mysterious symptoms, I went to an allergist who gave me tests. I was diagnosed with gluten intolerance. Overnight, I changed the way I eat. I completely altered my diet to remove all bread and gluten products. It was a new life! My digestion improved. I found incredible energy, lost weight, no more breathing problems. My skin became clear and stopped hurting. I have rediscovered my body and my vitality.

Modern Wheat Is Less Nutritious

Mineral content in modern wheat cultivars has significantly decreased, including copper, iron, magnesium, manganese, phosphorus, selenium, and zinc. High levels of these nutrients can be found in landraces and old low-yielding varieties. Because wheat landraces have been evolved in environments with low nutrient availability, they represent a source of variation for selection of varieties adapted to cropping systems with low fertilizer input. Compared to the cost associated with the formation of new roots, arbuscular mycorrhizae may considerably increase the active absorbing root surface with minor cost to the wheat plant, thus enhancing the uptake of phosphorus, in particular, and other macro- and micro-nutrients.[6]

Selection for high yield has caused a decrease in protein and mineral content and a loss of nutrition and flavor. Although there is an increase in the yield in high-input chemical systems, yield decreases in low-input organic systems. Shorter plants cannot compete with weeds. Stubby roots have lower nutrient uptake capacity. Weakened plants are bred with dependence on agrochemical protectants to survive. In pure line varieties diseases spread epidemically when all plants are the same. But this isn't the case if there is diversity within the population, because then there are varying resistances in the field.

The modern dependence on agrochemicals contributes to climate change by nitrogen leaching, greenhouse gas emissions, and pesticides. Although modern wheat varieties may have high yields in agrochemical systems, they have inferior baking quality and mineral content. To improve yields in organic wheat production, increasing diversity with wide spacing in organic soil is more important than introducing new pure line varieties.

The Broadbalk Wheat Experiment,[7] spanning more than a century, has documented the decline in the nutrition of modern wheat. Since 1843, generations of investigators have grown wheat in parallel plots. Each year the harvest was tested for nutritional content. The findings are alarming. Although yield suddenly increased in 1968 when dwarf wheats were introduced, the nutritional value plummeted 18 to 29 percent in mineral content of zinc, iron, copper, magnesium, phosphorus, manganese, sulfur, and calcium. We are eating an empty harvest, craving more calories to get the basic nutrition we need.

Less Is More

In a study conducted in the Czech Republic, landraces and old cultivars were shown to have a higher content of protein than modern wheats. An increase in grain yield tends to cause a decrease in protein content.[8]

Comparing modern to older varieties, the modern ones have a small, 10 to 15 percent, improvement in the yield/protein continuum, over 150 years of breeding. The great advantage of modern varieties is that under high nitrogen they do not lodge.

Monocrop or Polyculture Diet

The Green Revolution has changed dietary habits. The comparison between traditional systems of agriculture and Green Revolution agriculture often neglects the full picture. Green Revolution agriculture produces monocultures of cereal grains, while traditional agriculture incorporates nutritionally balanced polycultures of grain, legumes, and vegetables. Monocrops of modern wheat have replaced the traditional polyculture of wheat, legumes, and vegetables. Although fewer people die from starvation, more are affected by malnutrition such as iron or vitamin deficiencies. Modern wheat monocultures are often used for export or feed for animals. In contrast, emmer and einkorn are exceptionally high in health-promoting phytonutrients and antioxidants. Einkorn has high levels of lutein and beta-carotenes. Lutein, an antioxidant, reduces the risk of age-related macular degeneration (AMD), the leading cause of vision loss among older Americans, and provides protection against heart disease and cancer.[9] In North America, the daily intake of lutein is far below the recommended level and is declining due to a decrease in the consumption of the dark green vegetables that are its main source.

Celiac Stimulatory Epitopes

The worldwide rise in celiac disease occured at the exact same time as the introduction of Green Revolution dwarfed wheat. Dr. Hetty van den Broeck has long studied wheat and the gluten proteins that are responsible for celiac disease. Dr. van den Broeck is part of a research group in Holland that compared gluten proteins in modern wheats and in landrace wheats grown a century ago. The presence of the celiac disease stimulatory culprit, the Glia-α9 epitope, is alarmingly higher in modern wheat. Dr. van den Broeck's extensive analysis of scores of modern and heritage wheats conclusively proves that modern wheat is making us sick. Modern wheat breeding has increased in the protein parts that cause celiac disease, known as epitopes.[10] Norman Borlaug, the Nobel Prize–winning wheat breeder, not only introduced higher-yielding wheat but inadvertently created a high-gluten wheat that humans have not evolved to digest. Modern wheat is making us sick.[11]

According to Dr. van den Broeck:

> The immune system is stimulated by gluten proteins. Once the system is triggered, it alters the mucosa of the small intestine, which flattens gradually causing symptoms of diarrhea, malnutrition, and malabsorption of nutrients. Not all celiac patients have flattened mucosa. Celiac can occur without flattened mucosa, which is then called latent celiac disease. People get sick. Each patient is different. Some get sick at the sight of a morsel of bread, so to speak, whereas others may eat a pizza a week without major symptoms, or with the consequences they are able to manage.[12]

The hidden gluten in our diets is a serious problem. "Modern gluten is not only in bread, but it is hidden in many food products from soups, sauces, meat products, potato chips, candies, ice creams, even in medicines and vitamin supplements, because it has useful adhesive properties for processed foods. For example, gluten is added to potato chips to adhere flavors."

Dr. van den Broeck is trying to find solutions by identifying which species and varieties of wheat, ancient or modern, do not cause an allergic response:

> We are looking for wheat varieties that are low in celiac disease epitopes, so that they can be used by breeders to create new varieties of wheat with safer gluten for consumers. It is possible to reduce the symptoms, but also the frequency of celiac disease by identifying wheat with less toxic gluten.[13]

Resources

I t takes more than a book. Just as the seeds are alive, the resources, knowledge, and skills to restore landrace wheat and bake artisan bread can come alive as part of a dynamic community network.

It is my greatest hope that more and more people work together to restore the forgotten landrace wheats. Like Gandhi's Salt Walk that encouraged people to collect their own free salt from the sea, saving landrace wheat seed is a powerful way to restore our connection to the abundance of nature and to liberate ourselves from corporate control of our seed and food systems.

It takes a community to restore a landrace. It is most effective to work in teams so that a vital, diverse meta-population can be restored. Restoring the culture of seed saving is as important as the scientific aspects. A handful of landrace seed can become a field of abundance that nourishes local community partnerships.

It is important that anyone requesting seeds from a seed bank understand how rare the seeds are. A seed bank is not a seed company. Seed banks store treasures of threatened biodiversity that cannot be replaced. Let us trial, evaluate, and select each accession with the utmost stewardship. If you have questions, please contact me though my blog at www.growseed.org.

THE HERITAGE GRAIN CONSERVANCY

For up-to-date resources, events, landrace wheat varities, and a community blog to share questions and experiences, please visit www.growseed.org.

SEED BANKS

The U.S. seed bank Germplasm Resources Information Network (GRIN) maintains about 60,000 samples of diverse wheats collected throughout the world. To request seed from GRIN, please visit http://www.ars -grin.gov/npgs/searchgrin.html.

The CIMMYT Wheat Germplasm Bank, the largest wheat seed bank in the world, maintains nearly 100,000 accessions of wheat. CIMMYT disseminates wheat-breeding populations to evaluate and select for adaptability to local environments. Seed is freely shared on request to researchers, students, and institutions worldwide. For an overview of CIMMYT, please visit http://www.cimmyt.org. For the CIMMYT Wheat Germplasm Bank, please visit https://slate.adobe.com /cp/QXtG2/.

WHEAT AND SEED-SAVING PROGRAMS, CONFERENCES, AND COMMUNITY SEED BANKS

For the Northeast Grain Renaissance, please visit www.grownyc.org.

For the Maine Grain Alliance Kneading Conference, please visit www .kneadingconference.com.

For Sylvia Davatz's Solstice Seed Catalogue, please visit https://uvlocal vores.files.wordpress.com/2015/01/solsticeseedscatalogue2015.pdf.

For the Brockwell Bake Association, please visit www.brockwell-bake .org.uk.

For Prairie Garden Seeds, please visit www.prseeds.ca.

For Kokopelli, please visit www.kokopelli-seed-foundation.com

For England's heritage wheat see a profile of John Letts at www.bakerybits .co.uk/john-letts.

For German Biodynamic Wheat Breeding, please visit www.darzau.de.

For the French Peasant Seed Network, please visit www.semences paysannes.org.

For Greek landrace wheat, please visit www.aegilops.gr and www.peliti.gr.

For New Zealand heritage wheats, please visit https://heritagefoodcrops .org.nz/ancient-wheat.

For Seed Freedom, please visit http://seedfreedom.info.

For Navdanya ("Nine Seeds"), please visit www.navdanya.org.

For a Nepal case study on community seed banks, please visit http://www.bioversityinternational.org/uploads/tx_news/Community _seed_banks_in_Nepal__past__present_and_future_1642.pdf.

For the Sacred Seed Sanctuary, please visit http://sacredseedssanctuary.org.

For information on community seed banks, please visit bioversityinter national.org/fileadmin/user_upload/online_library/publications/pdfs /Community_Seed_banks/Community_Seed_Banks.pdf.

For Bread Therapy, please visit http://breadtherapy.net.
For Ecological Plant Breeding, please visit www.eco-pb.org.
For European seed savers, please visit www.liberatediversity.org.
For Charles Eisenstein's Sacred Economics, please visit www.sacred
-economics.com.
For "A Grounded Guide to Gluten" by Cornell University's Lisa Kucek,
please visit http://onlinelibrary.wiley.com/enhanced/doi/10.1111
/1541-4337.12129/.
For detailed explanations of biodynamic methods, please visit www.bio
dynamics.in.
For biodynamic preparations, please visit www.jpibiodynamics.org.

BOOKS

Carleton, Mark. *The Basis for the Improvement of American Wheats.*
Washington, D.C.: USDA, 1900. http://wholegrainconnection.org/site
buildercontent/sitebuilderfiles/carlton24basisforimprovem24carl.pdf.
Carleton, Mark. *The Small Grains.* New York: Macmillan, 1920. https://
books.google.com/books?id=MMk2AAAAMAAJ&printsec=frontcover
&dq=small+grains&lr=&ei=kEXYS4a3D4TAlQSF7ICLDw&cd
=1#v=onepage&q&f=false.
Clark, J. Allen. *Classification of America's Wheat Varieties.* Washington,
D.C.: USDA, 1922. http://growseed.org/1923%20WHEAT%20
HISTORY.pdf.
Todd, Sereno Edwards. *American Wheat Culturalist,* New York: Taintor
Brothers, 1868.

PROCEEDINGS

For a record of proceedings of the first International Workshop on Hulled
Wheats (Tuscany, Italy, 1995) please visit https://www.bioversity
international.org/index.php?id=244&tx_news_pi1[news]=148&
cHash=7de9d50f97676192303fc08a19002211.
For restoring ancient wheat in Israel, Palestine, and Jordan, please see:
Proceedings: http://igb.agri.gov.il/main/index.pl?page=112.
Project Report: http://growseed.org/IPSC.pdf.

ON-FARM CONSERVATION
AND PARTICIPATORY BREEDING

For European Landraces On-Farm Conservation, please visit http://www
.bioversityinternational.org/fileadmin/_migrated/uploads/tx_news
/European_landraces__on-farm_conservation__management_and
_use_1347.pdf.

For "Evolutionary Populations: Living Gene Banks in Farmers' Fields,"
please visit http://www.agriculturesnetwork.org/magazines/global
/cultivating-diversity/plant-breeding.

For Wheat Landraces: Genetic Resources for Sustenance and
Sustainability (Abdullah Jaradat) please visit http://www.ars.usda.gov
/SP2UserFiles/Place/50600000/products-wheat/AAJ-Wheat
%20Landraces.pdf.

FILMS AND VIDEOS

For "La Passion du Pain" by Nicolas Supiot, my teacher, please visit
https://www.youtube.com/watch?v=8B_7AFYmkYo.

For Jean Francois Berthelot, French seed saver extraordinaire, please visit
https://www.youtube.com/watch?v=vxgOycrG0GY.

For *The Story of Seeds*, please visit www.opensesamemovie.com.

Notes

Chapter 1: On the Verge of Extinction

1. A "landrace," the first stage in domestication of wild food crops, contains a genetically diverse population of plants or animals that are uniquely well adapted to their local environment.
2. Salil Singh, "Norman Borlaug: A Billion Lives Saved," AgBioWorld, accessed February 9, 2016, http://www.agbioworld.org/biotech-info/topics/borlaug/special.html. See also Vandana Shiva, *Violence of Green Revolution: Third World Agriculture, Ecology, and Politics* (Lexington, KY: University Press of Kentucky, 2016).
3. "What Is Happening to Agrobiodiversity?," Food and Agriculture Organization of the United Nations, accessed 9 February 2016, http://www.fao.org/docrep/007/y5609e/y5609e02.htm.
4. Martha H. Hamilton, *The Great American Grain Robbery and Other Stories* (Washington: Agribusiness Accountability Project, 1972). See also Vandana Shiva, "The Great Grain Robbery by Agribusiness MNC's," ZNet, published June 21, 2006; José Ciro Martínez, "Give Us This Day Our Daily Bread? The Politics of Flour in Hashemite Jordan," Jadaliyya, published January 28, 2014; Hans Herren, *The Great Food Robbery: How Corporations Control Food, Grab Land, and Destroy the Climate* (Oxford: Pambazuka Press, 2012); Andrew Bosworth, "Mutant Seeds for Mesopotamia," InfoWars, published October 16, 2008; and Norman J. Church, "Why Our Food Is So Dependent on Oil," *Resilience*, April 1, 2005.
5. J. M. Cooper, G. Butler, and C. Leifert, "Life cycle analysis of greenhouse gas emissions from organic and conventional food production with and without bio-energy options," *NJAS – Wageningen Journal of Life Sciences* 58, no. 3–4 (2011): 185–92, http://sciencedirect.com/science/article/pii/S1573521411000340.
6. Peter Melchett, "Glyphosate Scientific Briefing," Soil Association, http://www.senseaboutscience.org/data/files/VoYS/SOIL-ASSOCIATION_SLIDES_glyphosate_scientific_briefing.pdf.

7. David Saltmiras, "Ask Us Anything About GMOs!," GMOAnswers, published August 20, 2013, http://gmoanswers.com/ask/if-roundup -safe-human-consumption-trace-amounts-food-then-it-safe-drink-it -if-not-where-line.

8. "The truth about Roundup and wheat: support material," Kansas Wheat, http://kswheat.com/the-truth-about-roundup-and-wheat -support-material; See also "Preharvest Staging Guide," Monsanto, http://roundup.ca/_uploads/documents/MON-Preharvest %20Staging%20Guide.pdf.

9. "Adult Obesity," Harvard T. H. Chan, http://www.hsph.harvard.edu /obesity-prevention-source/obesity-trends/obesity-rates-worldwide /index.html.

10. Sarah J. Hale, *The Good Housekeeper* (Boston: Weeks, Jordan and Company, 1839).

11. "Barm" is the sediment that sinks to the bottom of fermented alcoholic beverages such as beer or wine. It was used to leaven bread or as a starter to ferment in a new batch. For instructions see Monica Spiller, "Introducing *Barm* Bread 2010," Barm Bread, published September 3, 2010, sustainablegrains.org/sitebuildercontent /sitebuilderfiles/barmbreadintroduction090310.pdf.

12. Harvey W. Wiley, *Foods and Their Adulteration* (Philadelphia: P. Blakiston's Son and Company, 1911).

13. Methyl cellulose is a cellulose-derived powder that is not digest-ible but is nontoxic. When dissolved in water it forms a gel commonly used as a thickener and in cosmetic products.

14. Personal conversation.

15. See Jean-François Berthelot, *Culture de blé bio*, video, 6:31, March 30, 2011, http://www.youtube.com/watch?v=vxgOycrG0GY.

16. Gluten is a "continuous proteinaceous matrix in the mature dry grain derived from the nitrogen absorbed from the soil. This is brought together to form a viscoelastic network when flour is mixed with water to form dough." Peter R. Shewry et al., "The Structure and Properties of Gluten: An Elastic Protein from Wheat Grain," *Philosophical Transactions of the Royal Society*, Biological Sciences 357, no. 1418 (2002): 133–142, http://www.ncbi.nlm.nih.gov/pmc /articles/PMC1692935/.

17. Lisa Kissing Kucek et al., "A Grounded Guide to Gluten: How Modern Genotypes and Processing Impact Wheat Sensitivity," *Comprehensive Reviews in Food Science and Food Safety* 14, issue 3 (May 2015): 285–302, http://onlinelibrary.wiley.com/enhanced /doi/10.1111/1541-4337.12129.

18. Karl-Josef Mueller, personal communication with author.

19. S. Bakhøj et al., "Lower Glucose-Dependent Insulinotropic Polypeptide (GIP) Response but Similar Glucagon-Like Peptide 1 (GLP-1), Glycaemic, and Insulinaemic Response to Ancient Wheat Compared to Modern Wheat Depends on Processing," *European Journal of Clinical Nutrition* 57 (2003): 1,254–1,261, http://www .nature.com/ejcn/journal/v57/n10/abs/1601680a.html.

Chapter 2: Forgotten Grains

1. See Restoring Ancient Wheat conference proceedings at http:// growseed.org/wheat.html.

2. Our group continues to meet in gatherings hosted by partner countries throughout the European Union to this day. We established a working group for landrace wheat conservation to exchange seed and knowledge. For more information, visit the Let's Liberate Diversity! website, www.liberatediversity.org.

3. Charles Mann, "Birth of Religion," *National Geographic*, June 2011.

4. Frank Morton, "Seed in a Whole Farm System," Heritage Grain Conservancy, http://growseed.org/frank.html. Frank Morton is an ecological plant breeder. See his website at www.wildgardenseeds.com.

5. For more on the geographic origins of crops, see Paul Gepts, "Where Did Agriculture Start?," University of California, Davis, http://www .plantsciences.ucdavis.edu/gepts/pb143/LEC10/Pb143l10.htm.

6. J. R. Harlan, "Agricultural Origins: Centers and Non-Centers," *Science* 174, no. 4008 (1971): 468–74; see also J. R. Harlan, "Our Vanishing Genetic Resources," *Science* 188, No. 4188 (1975): 618–621.

7. Abdullah A. Jaradat, "The value of wheat landraces," *Emirates Journal of Food and Agriculture* 26.2 (2014).

8. For more information see "Taxonomy of Wheat," Wikipedia, https:// en.wikipedia.org/wiki/Taxonomy_of_wheat.

9. Known as "Dika" in Georgia.

10. R. E. Oliver et al., "Evaluation of Fusarium Head Blight Resistance in Tetraploid Wheat (*Triticum turgidum* L.)," *Crop Science Society of America* Vol. 48, no. 1 (2008).

11. Hildegard von Bingen, *Hildegard von Bingen's Physica: The Complete English Translation of Her Classic Work on Health and Healing*, trans. Priscilla Throop (Rochester, VT: Healing Arts Press, 1998).

12. To learn more about Georgia's fascinating cuisine and folklore, see Darra Goldstein, *Georgian Feast: The Vibrant Culture and Savory Food of the Republic of Georgia* (Oakland: University of California Press, 2013); and Michael Berman, *Georgia Through Its Folktales* (London: Moon Books, 2010).

13. The vast collections of landrace seeds stored in world gene banks are available only in tiny amounts of one gram (20 seeds) to five grams (100 seeds).

14. Carol Deppe, *Breed Your Own Vegetable Varieties* (White River Junction, VT: Chelsea Green Publishing, 1993).

15. John H. Martin, *Polish and Poulard Wheats*, USDA, published June 1923, http://plantbreeding.wsu.edu/1923PolishAndPoulard Wheats.pdf.

16. Dr. Mark Hutton, "On-Farm Seed-Saving Workshop," Presentation at Restoring Our Seed Conference, MOFGA, 2008.

17. Brett Grohsgal, "On-Farm Seed-Saving Workshop," Presentation at Restoring Our Seed Conference, MOFGA, 2008.

18. Composite-cross gene pools are varietal mixtures crossed together to increase the diversity of a population.

19. Abdullah A. Jaradat, "The value of wheat landraces," *Emirates Journal of Food and Agriculture* 26, no. 2 (2014).

20. Raoul Robinson, *Return to Resistance* (Davis, CA: agAccess, 1995).

21. For a step-by-step instructional guide with photographs of how to hand-pollinate wheat, see http://growseed.org/pollinating-wheat.pdf.

22. Anders Borgen, communication with the author.

23. Adapted from S. Edwards Todd, *The American Wheat Culturist: A Practical Treatise of the Culture of Wheat* (New York: Taintor Brothers & Co., 1868).

24. Plant breeding is especially engaging for fifth- and sixth-graders in their golden age prior to puberty, to develop a deeper

appreciation of "the birds and the bees" than our media-saturated culture provides.

25. Raoul Robinson, *Return to Resistance* (Davis, CA: agAccess, 1995).
26. Raoul Robinson, "Horizontal Resistance," Heritage Grain Conservancy, http://growseed.org/HR.html.

CHAPTER 3: LANDRACE GRAIN HUSBANDRY

1. Dr. Alan Betts, "Climate Change in Vermont," October 29, 2011, http://alanbetts.com/workspace/uploads/vtccadaptclimatechangevtbetts-1323873554.pdf.
2. J. F. Schafer, "Rusts, Smuts and Powdery Mildew," in *Wheat and Wheat Improvement,* 2nd ed. (Madison, WI: American Society of Agronomy, 1987), 542–84.
3. William Albrecht, "The Loss of Organic Matter and Its Restoration," in *Soils and Men: USDA Yearbook of Agriculture* (Washington, D.C.: USDA, 1938).
4. J. A. Kirkegaard et al., "Suppression of Soil-Borne Cereal Pathogens and Inhibition of Wheat Germination by Mustard Seed Meal" (Proceedings of the 8th Australian Agronomy Conference, Queensland, Australia, January 30–February 2, 1998).
5. Carlos Perez, "Use of Green Manures to Reduce Inoculum Production of *Fusarium graminearum* on Wheat Residues," SARE: Sustainable Agriculture Research & Education, http://mysare.sare.org/sare_project/gnc05-054/?page=final.
6. CR Lawn, a master organic farmer and founder of Fedco Seed Company, is my research partner in the on-farm heritage wheat trials.
7. Rudolf Steiner, "Agriculture Course: Lecture 4," Rudolf Steiner Archive and e.Lib., published June 26, 2007, http://wn.rsarchive.org/Lectures/GA327/English/BDA1958/19240612p01.html.
8. Ghazi Al-Karaki et al., "Field Response of Wheat to Arbuscular Mycorrhizal Fungi and Drought Stress," *Mycorrhiza* 14 (2004): 263–69, http://mycorrhiza.ag.utk.edu/h2o/alkaraki_2004_mycorrhiza.pdf.
9. M. A. Monreal et al., "Crop Management Effect on Arbuscular Mycorrhizae and Root Growth of Flax," *Canadian Journal of Plant Science* 91, no. 2 (2011): 315–324, http://pubs.aic.ca/doi/abs/10.4141/CJPS10136.

10. Cemal Yucel et al., "Screening of Wild Emmer Wheat Accessions for Mycorrhizal Dependency," *Turkish Journal of Agriculture and Forestry* 33 (2009): 513–523, http://journals.tubitak.gov.tr/agriculture/issues /tar-09-33-5/tar-33-5-11-0902-47.pdf.

11. Gary Nabhan, *Where Our Food Comes From: Retracing Vavilov's Quest to End Famine* (Washington, DC: Island Press, 2008).

12. "Cultivar Mixtures, Cover Crops, and Intercropping with Organic Spring Wheat," University of Manitoba Natural Systems Agriculture, published June 2005, http://www.umanitoba.ca/outreach /naturalagriculture/articles/wheatintercrop.html.

13. Vilmorin-Andrieux et cie., *Les Meilleurs Blés: Description et culture des principales variétés de froments d'hiver et de Printemps* (Ann Arbor: University of Michigan Library, 1880).

14. Persian a.k.a. Dika is *T. carthlicum.*

15. M. Jorjadze et al., "The Ancient Wheats of Georgia and Their Traditional Use in the Southern Parts of the Country," *Emirates Journal of Food and Agriculture* 26, no. 2 (2014): 192–202, http://ejfa .me/?mno=185903.

16. S. Edwards Todd, *The American Wheat Culturist–A Practical Treatise on the Culture of Wheat* (New York: Taintor Brothers and Company, 1868).

17. For more information, see http://sri.cals.cornell.edu.

18. Frederick Watts, "Tests of the Department of Seed," in *Annual Reports of the Department of Agriculture* (1871).

19. S. Edwards Todd, *The American Wheat Culturalist–A Practical Treatise on the Culture of Wheat* (New York: Taintor Brothers and Company, 1868).

20. Vilmorin-Andrieux et cie., *Les Meilleurs Blés: Description et culture des principales variétés de froments d'hiver et de Printemps* (University of Michigan Library, 1880).

21. Ibid.

22. Glenn Roberts, communication with the author.

23. Sharon Palmer, "Digging Into Soil Health," *Today's Dietitian* 11, no. 7 (July 2009): 38, http://www.todaysdietitian.com/newarchives /062909p38.shtml.

24. Humates contain 65 percent humic acid in a carbon matrix to chelate minerals and nutrients, making them more easily available

to soil microorganisms and plants. Extensive tests show marked crop improvement when Menefee Humates were applied. Menefee Humates are available through http://fedcoseeds.com. Azomite, known to Native Americans for its "healing power," contains 70 essential minerals and trace elements in a balanced ratio. The cation exchange capacity (CEC) is a measure of fertility, the nutrient retention capacity of soil. One way to increase CEC is to increase the formation of humus.

25. Ehrenfried Pfeiffer, *Weeds and What They Tell* (Edinburgh, UK: Floris Books, 2012).

26. Leaf-tissue testing involves taking samples from the plant and sending them to a laboratory for mineral nutrient analysis. Sap testing involves taking leaf petioles and expressing the sap, which is tested for nitrate and/or potassium using portable meters. See G. Hochmuth, et al., "Plant Tissue Analysis and Interpretation for Vegetable Crops in Florida," University of Florida IFAS Extension, published January 1, 1991, http://edis.ifas.ufl.edu/ep081; and "Plant Testing," Pike Agri-Lab Supplies, Inc., http://www.pikeagri.com/products/plant-testing-meters.

27. Horn silica spray contains finely ground quartz crystals placed in a cow's horn and buried during the hot season in early April, then lifted out in September and stirred. It is stored in a glass jar on a sunny windowsill. See "Preparation 501 - Cow Horn Silica," Bio-Dynamic Association of India (BDAI), http://biodynamics.in/BD501.htm.

28. Sil-MATRIX is available from "Agriculture and Animal Feed," PQ Corporation, http://www.pqcorp.com/pc/North-America/Markets/Agriculture-and-Animal-Feed. Fulvic acid, a by-product of decomposition, increases enzyme activity and causes cell membranes to become more permeable. Water is able to enter cells at a higher rate, promoting balanced hydration and allowing the organism to pass unwanted toxic substances.

29. Edward Faulkner, *Plowman's Folly* (Norman: University of Oklahoma Press, 2012).

30. Masanobu Fukuoka, *The One-Straw Revolution*, trans. Larry Korn (New York: New York Review Books Classics, 2009).

31. Larry Korn, "Natural Farming of Fukuoka," *Mother Earth News* 52 (August 1978).

32. Masanobu Fukuoka, *The One-Straw Revolution*, trans. Larry Korn (New York: New York Review Books Classics, 2009).

33. "Wheat Disease Identification," University of Wisconsin–Madison Cooperative Extension, http://fyi.uwex.edu /fieldcroppathology/files/2010/11/Wheat_Disease_ID.pdf.

34. Klaas Martens, communication with the author.

35. Rudolf Steiner, "Agriculture Course, Lecture 2," Rudolf Steiner Archive and e.Lib., published June 26, 2007, wn.rsarchive.org /Lectures/GA327/English/BDA1958/19240610p01.html.

36. "Applied Biodynamics Newsletter (Winter 2009–2010)," Josephine Porter Institute, https://jpibiodynamics.org/wp-content /uploads/2014/03/Equisetum-BD508issue.pdf.

37. Ibid.

38. Mary-Howell Martens and Klaas Martens, "Harvesting High Quality Organic Grain," *Acres USA* 32, no. 10 (October 2002).

39. Shmurah ("watched, guarded") Matzah (Hebrew: מַצָּה שְׁמוּרָה) is made from grain that has been under constant supervision from the moment of harvest to ensure no contact with moisture so that no fermentation can occur.

40. Klaas Martens, communication with the author.

41. Christine Karutz, "Ecological Cereal Breeding and Genetic Engineering," Research Institute for Organic Agriculture (FiBL), published April 1998, http://orgprints.org/4855/1/karutz.htm.

42. S. Edwards Todd, *The American Wheat Culturist* (New York: Taintor Brothers and Company, 1868).

43. "Farmer Built Spelt Dehuller," SARE: Sustainable Agriculture Research & Education, http://mysare.sare.org/sare_project /FNE11-731/.

44. "Potential boost for world's food supply: Resistance gene found against Ug99 wheat stem rust pathogen," Science Daily, published June 27, 2013, https://www.sciencedaily.com/releases/2013/06 /130627141726.htm.

45. Rudy Ruitenberg, "Australian Salt-Tolerant Wheat May Aid Food Security," Bloomberg Business, published March 12, 2012, http:// www.bloomberg.com/news/articles/2012-03-12/salt-tolerant-wheat -developed-in-australia-may-aid-food-security.

46. Daniel Zohary and Maria Hopf, *Domestication of Plants in the Old World: The Origin and Spread of Cultivated Plants in West Asia, Europe, and the Nile Valley* (London: Oxford University Press, 2001).
47. Mária Hajnalová and Dagmar Dreslerová, "Ethnobotany of einkorn and emmer in Romania and Slovakia: towards interpretation of archaeological evidence," *Památky Archeologické* CI, 169-202.

Chapter 4: Journey of the Sheaves: Grain Folk Traditions

1. Personal communication.
2. Transcription by author of excerpt from *The World of the Goddess – Marija Gimbutas*, video, 1:42:56, December 6, 2011, https://www.youtube.com/watch?v=yU1bEmq_pf0.
3. Triptolemus, a mythic archetypal man in the Eleusinian mysteries, was the inventor of the plow and agriculture. In Homer's hymn on Demeter, Triptolemus was a noble man instructed by Demeter in sacred worship.
4. Carl Kerényi, *Eleusis: Archetypal Images of Mother and Daughter*, trans. Ralph Manheim (Princeton, NJ: Princeton University Press, 1991).
5. Jack R. Harlan, *Crops & Man* (Madison, WI: American Society of Agronomy, Crop Science Society, 1992).
6. Anthony Christie, *Chinese Mythology* (Middlesex, England: Hamlyn Publishing Group, Ltd., 1968).
7. Carol Clark-Emory, "Ancient Beer Brewed to Include Antibiotic," Futurity, published September 2, 2010, http://www.futurity.org/ancient-beer-brewed-to-include-antibiotic/.
8. Miguel Civil, "Ninkasi, the Sumerian Goddess of Brewing and Beer," Beer Advocate, published December 20, 2000, http://www.beeradvocate.com/articles/304. Another translation of the hymn can be seen at "A hymn to Ninkasi: translation," The Electronic Text Corpus of Sumerian Literature, http://etcsl.orinst.ox.ac.uk/.
9. Stan Gooch, *Dream Culture of Neanderthals: Guardians of Ancient Wisdom* (Rochester, VT: Inner Traditions, 2006).
10. Stan Gooch, *Cities of Dreams: When Women Ruled the Earth* (London: Aulis Publishers, 1995). See Stan Gooch, "Excerpts from *Cities of Dreams*," *AULIS* Online, 1995, http://www.aulis.com/twothirds7.htm.

11. "New Evidence Challenges Old Assumptions About Gobekli
Tepe," SF-Fandom's History Blog, published August 24, 2012, http://
history.sf-fandom.com/2012/08/24/new-evidence-challenges
-old-assumptions. Einkorn was also found at Tell Abu Hureyra in
modern-day Syria and Jericho circa 11,500 years ago (9500 BCE).
12. Ibid.
13. Jack R. Harlan, "A Wild Wheat Harvest in Turkey," *Archaeology*
20 (1967): 197–201.
14. Raanan Tzarfati et al., "Threshing Efficiency as an Incentive for
Rapid Domestication of Emmer Wheat," *Annals of Botany* (2013),
http://aob.oxfordjournals.org/content/early/2013/07/24
/aob.mct148.full.
15. Ehud Weiss et al., "Small-Grained Wild Grasses as Staple Food
at the 23,000-Year-Old Site of Ohalo II, Israel," *Economic Botany* 58
(Winter 2004): S125–S134, http://www.jstor.org/stable/4256914.
16. Celtic descendants in Europe included the Iron Age Hallstatt
culture (c. 800–450 BCE) in Austria, the British Isles, France,
Bohemia, Poland, most of Central Europe, the Iberian Peninsula, and
northern Italy. Following the Gallic invasion of the Balkans in 279
BCE, Celts were as far east as central Anatolia, Turkey — before the
Roman Conquest. See *The Celts: History, Life, and Culture*, ed. John T.
Koch and Antone Minard (Santa Barbara, CA: ABC-Clio, 2012). See
also Fritz Zimmerman, "Early Celtic Art," Celtic Ruins, published
March 25, 2013, http://celticruins.blogspot.com/2013/03
/early-celtic-art.html.
17. Marija Gimbutas, *The Civilization of the Goddess: The World of Old
Europe* (New York: HarperCollins, 1991).
18. Eliso Kvavadze et al., "30,000-Year-Old Wild Flax Fibers," *Science*
325, no. 5946 (11 September 2009): 1,359, http://sciencemag.org
/content/325/5946/1359.short.
19. Karinė Khristoforovna Kushnareva, *The Southern Caucasus in
Prehistory: Stages of Cultural and Socioeconomic Development from
the Eighth to the Second Millennium B.C.*, trans. H. N. Michael
(Philadelphia: University of Pennsylvania Museum of Archaeology
and Anthropology, 1997).

20. S. A. Greg, *Foragers and Farmers: Population Interaction and Agricultural Expansion in Prehistoric Europe* (Chicago: University of Chicago Press, 1988).
21. Wolfgang Haak, "DNA Reveals Origins of First European Farmers," The University of Adelaide, November 10, 2010, http:// adelaide.edu.au/news/news42161.html.
22. William Ryan and Walter Pitman, *Noah's Flood: New Scientific Discoveries about the Event that Changed History* (New York: Simon and Schuster, 2000).
23. The Acheulian Goddess figurine, carved from stone, was found in modern-day Golan Heights in Israel and has been carbon-dated from 232,000 to 800,000 years ago. It was fashioned by a nomadic tribe of early humans who predate the Neanderthal period. According to the *Israel Journal of Prehistoric Society,* it is the earliest known manifestation of a work of art. See Naama Goren-Inbar, "A Figurine from the Acheulian Site of Berekhat Ram," *Mitekufat Haeven: Journal of the Israel Prehistoric Society* (1986): 7–12, http://www.jstor.org/stable/23373142 ?seq=1#page_scan_tab_contents.
24. Marija Gimbutas, *The Civilization of the Goddess* (New York: HarperCollins, 1991).
25. Fritjof Capra, *The Systems View of Life: A Unifying View* (Cambridge: Cambridge University Press, 2014); and Fritjof Capra, *The Hidden Connections* (New York: Anchor Books, 1989).
26. David Bohm, *Wholeness and the Implicate Order* (London: Routledge, 1980).
27. Ruth Tringham, "Archeology: The Civilization of the Goddess: The World of Old Europe," *American Anthropologist* 95, no. 1 (March 1993): 196–97, http://onlinelibrary.wiley.com/doi/10.1525/aa.1993 .95.1.02a00510/abstract.
28. Harald Haarmann and Joan Marler, "An Introduction to the Study of the Danube Script," *Journal of Archaeomythology* 4 (Winter 2008), http://archaeomythology.org/wp-content/uploads/2012 /01/2008-vol4-intro1.pdf.
29. Ruth Tringham, "Archeology: The Civilization of the Goddess: The World of Old Europe," *American Anthropologist* 95, no. 1 (March

1993) 196–97, http://onlinelibrary.wiley.com/doi/10.1525/aa.1993
.95.1.02a00510/abstract.

30. Marija Gimbutas, *Neolithic Macedonia as Reflected by Excavation at Anza, Southeast Yugoslavia* (Los Angeles: University of California Institute of Archaeology, 1976).

31. Marija Gimbutas, *Goddesses and Gods of Old Europe 6500–3500 B.C.* (London: Thames & Hudson, 1982).

32. Alex Imreh, "Th (sic) first Great Civilizations of Europe: Cucuteni–Trypillia 5000–3000 BC + Vinca 6-3millennium BC = I2a in Haplogroup I," published January 17, 2011, https://aleximreh .wordpress.com/2011/01/17/th-first-great-civilization-of-europe -cucutenitrypillia-50003000-b-c/; and Kantzveldt, "7,000 Year Old Origins of 'The Supreme Ultimate,'" Above Top Secret, published October 10 2012, http://www.abovetopsecret.com/forum /thread889324/pg1&mem=Kantzveldt.

33. James George Frazer, *The Golden Bough: A Study of Magic and Religion* (Oxford: Oxford University Press).

34. For further information, see Alexander Thom, *Megalithic Sites in Britain* (Oxford: Oxford University Press, 1967).

35. Máire MacNeill, *The Festival of Lughnasa* (Dublin: University College, 2008).

36. Ibid.

37. David W. Anthony, *The Horse, the Wheel, and Language: How Bronze-Age Riders from the Eurasian Steppes Shaped the Modern World* (Princeton, NJ: Princeton University Press, 2007).

38. Porus Homi Havewala, *The Saga of the Zoroastrian Race* (Mumbai, India: Arktos Press, 2011).

39. K. E. Eduljee, "Legend of the Grain of Wisdom," Heritage Institute, http://heritageinstitute.com/zoroastrianism/grain /index.htm.

40. Porus Homi Havewala, communication with the author. Havewala is a Zoroastrian historian and author of *The Saga*, an epic historic novel based in Zoroastrian traditions that spans tales of their origins 20,000 years ago before the Ice Age with Zoroastrian teachings.

41. Jacob Neusner, *The Fathers According to Rabbi Nathan* (Gainesville, FL: University of South Florida Press, 1986).

42. Adapted by author from Yehudah Steinberg, *Hebrew Stories for Children 1880–1890,* vol. 1, *Ba-Golah* (Jerusalem: Israel Press, 1944).

43. Rabbi Moses Maimonides, *Moreh Nevuchim Guide for the Perplexed,* trans. Dr. M. Friedlander (New York: E. P. Dutton and Company, 1904).

44. "Shmurah" is the Hebrew word for "watched" or "guarded" and refers to grain that is watched and protected from contact with moisture so that it can be used for kosher matzah.

45. Raphael Patai, *The Hebrew Goddess* (Detroit, MI: Wayne State University, 1978); and Carol Meyers, *Households and Holiness: The Religious Culture of Israelite Women* (Minneapolis, MN: Fortress Press, 2005).

46. Judith M. Hadley, *Cult of Asherah in Ancient Israel and Judah: Evidence for a Hebrew Goddess* (Cambridge: University of Cambridge Oriental Publications, 2000).

47. Talmud, Yoma 54b.

48. M. Harris, *Marriage as Metaphysics: A Study of the Iggeret HaKodesh by Moses Maimonides* (Cincinnati, OH: Hebrew Union College Annual, 1962).

49. Moshe Idels, *Kabbalah: New Perspectives* (New Haven, CT: Yale University Press, 2001).

50. The Hebrew and Arabic words *Beit Lehem* mean "Bread House."

51. For further information, see "Hestia the Greek Goddess," Chapel of Our Mother God, http://mother-god.com/hestia-the-greek-goddess.html.

52. Miguelonne Toussaint-Samat, *The History of Food,* trans. Anthea Bell (Cambridge, MA: Wiley-Blackwell, 2008).

53. For further information, see "Project – Wheat Dress," Pinterest, https://www.pinterest.com/francadonato/project-wheat-dress.

54. A. M. Watson, *Agricultural Innovation in the Early Islamic World* (London: Cambridge University Press, 1983).

55. Adapted from "Ibn al-'Awwām, Kitāb al-filāḥa," The Filaha Texts Project, http://www.filaha.org/author_ibn_al_awwam.html.

56. Gabriel Alonso de Herrera, *Ancient Agriculture,* trans. Rosa Lopez-Gaston (Santa Fe, NM: Ancient City Press, 2006).

57. Bridget Ann Henisch, *Fast and Feast: Food in Medieval Society* (University Park: The Pennsylvania State Press, 1976).

58. Thomas Tusser, *Five Hundred Points of Good Husbandrie* (London: English Dialect Society, Trubner, 1580).

59. John Letts, personal correspondence with the author. For more information on John Letts work visit: http://www.bbc.co.uk /insideout/southwest/series5/thatch.shtml; http://ihbc.org.uk /context_archive/56/thatchireland/nireland.html; and http:// bakerybits.co.uk/john-letts.

60. Emmanuel Le Roy Ladurie, *Times of Feast, Times of Famine: A History of Climate Since the Year 1000* (Garden City, NY: Doubleday, 1971).

61. Thomas Tusser, *Five Hundred Points of Good Husbandrie* (London: Trubner and Co., 1878 edition).

62. Keith Stavely and Kathleen Fitzgerald, *America's Founding Food: The Story of New England Cooking* (Chapel Hill: University of North Carolina Press, 2015).

63. B. Skovmand et al., "Evaluation of Oaxacan Wheat Landraces under Moisture Stress (C08-skovmand172009-Poster)," http:// download.clib.psu.ac.th/datawebclib/e_resource/e_database /agronomy/2002/Browse/pdf/C08-skovmand172009-Poster.pdf.

64. J. Allen. Clark, "Classification of American Wheat Varieties," in *USDA Bulletin No. 1074* (Washington, DC: USDA, 1922).

65. For an overview of European wheat breeding, see B. Belderok et al., *Bread-Making Quality of Wheat: A Century of Breeding in Europe* (Dordrecht, Netherlands: Kluwer Academic Publishers, 2000).

66. Nikolai Vavilov, *Scientific Basis for Wheat Breeding*, Vol. 13 (Waltham, MA: Chronica Botanica, 1950).

67. "Wadi Fukin," Heritage Grain Conservancy, http://growseed. org/wadifukin.html.

68. Marija Gimbutas, *Civilization of the Goddess* (San Francisco: Harper, 1991).

69. "Table Des Matiere," Les Meilleurs Blés, http://museum.agropolis .fr/pages/documents/bles_vilmorin/tome1/1_table_des_matieres1 .htm; and "Les conférences du Salon du bien-être et du bio," Ouest France, http://www.ouest-france.fr/les-conferences-du-salon-du -bien-etre-et-du-bio-772175.

70. John Ray, *Catalogue of Cambridge Plants*, trans. and ed. P. H. Oswald and Christopher David Preston (London: Ray Society, 2011).

71. William Ellis, *The Modern Husbandman, or, the Practice of Farming* (London: 1784).

72. "Blue Cone Rivet," BBA Wheat Portal, http://brockwell-bake.org .uk/wheat/hub.php?ID=51.

73. "Chapter Fourteen: Spread into the Kansas Valley Counties," *Kansas History*, http://www.kansashistory.us/fordco/malin /14.html.

74. J. Allen Clark, John H. Martin, and Carleton R. Ball, *Classification of American Wheat Varieties*, United States Department of Agriculture Bulletin No. 1074 (1922): https://archive.org/details/classification of1074clar.

75. Mark A. Carleton, "Successful Wheat Growing" in the *USDA Yearbook of Agriculture 1900*, 529–542, http://naldc.nal.usda.gov /naldc/catalog.xhtml?id=IND43620518.

76. "From a Single Seed - Tracing the Marquis Wheat Success Story in Canada to Its Roots in the Ukraine (6 of 11)," Agriculture and Agri-Food Canada, http://agr.gc.ca/eng/news/science-of-agricultural -innovation/from-a-single-seed-tracing-the-marquis-wheat -success-story-in-canada-to-its-roots-in-the-ukraine-6of11/?id =1181307781375.

77. *Maine Agriculture Society Report of 1857* (Augusta: Maine Board of Agriculture, 1837).

78. Ibid.

79. "Taxon: *Triticum aestivum* L. subsp. *aestivum*," U.S. National Plant Germplasm System, https://npgsweb.ars-grin.gov/gringlobal /taxonomydetail.aspx?id=40544.

80. Cyrus Gurnsey Pringle, *Record of A Quaker Conscience* (New York: Macmillan, 1908).

81. Henry Jackson Waters, *Missouri Agricultural Experiment Station*, Bulletin 15, July 1891.

82. "Germplasm Resources Information Network," USDA Agricultural Research Service, http://www.ars-grin.gov/.

83. Nikolai Vavilov, *The Origin, Variation, Immunity and Breeding of Cultivated Plants*, trans. K. Starr Chester, Ph.D., *Chronica Botanica* 13, no. 6/1 (1951): http://krishikosh.egranth.ac.in/bit-stream/1/2037885/1/P2466.pdf.

Chapter 5: A Taste of History

1. Regula Steinhauser-Zimmerman, Irmgard Bauer, and Sabine Karg, *A Culinary Journey through Time* (Brøndby Strand, Denmark: Communicating Culture, 2011).
2. John Roach, "'Antibiotic' Beer Gave Ancient Africans Health Buzz," *National Geographic*, published May 16, 2005, http://news.national geographic.com/news/2005/05/0516_050516_ancientbeer.html.
3. "The Code of Hammurabi," trans. By L. W. King, Yale Law School Lillian Goldman Law Library, The Avalon Project, http://avalon.law .yale.edu/ancient/hamframe.asp.
4. The Talmud is the dynamic central text of Rabbinic Judaism. It contains the teachings of thousands of rabbis (from the Second Temple period of Israel to the present day) on Jewish ethics, law, philosophy, customs, history, medicine, folklore, and more. The Talmud is the basis for all codes of Jewish law and is added to each generation with contemporary commentaries.
5. Babylonian Talmud, Pesach. 107a. See "The Babylonian Talmud — Pesachim," Judeo-Christian Research, 2009, http://juchre.org /talmud/pesachim/pesachim.htm.
6. Babylonian Talmud. B. Bat 96b. See "The Babylonian Talmud — Baba Bathra," Judeo-Christian Research, 2009, http://juchre.org/talmud /bababathra/bababathra.htm.
7. G. Famularo et al., "Probiotic Lactobacilli: An Innovative Tool to Correct the Malabsorption Syndrome of Vegetarians?," *Medical Hypotheses* 65, no. 6 (2005): 1132–35, http://www.ncbi.nlm.nih.gov/ pubmed/16095846.
8. Stephen Harrod Buhner, *Sacred and Herbal Healing Beers* (Boulder, CO: Siris Books, 1998).
9. *The Domostroi: Rules for Russian Households in the Time of Ivan the Terrible*, ed. Carolyn Johnston Pouncy (New York: Cornell University Press, 1994).
10. R. E. F. Smith and David Christian, *Bread and Salt: A Social and Economic History of Food and Drink in Russia* (Cambridge: Cambridge University Press, 1984).
11. Ibid.

12. *Curye on Inglysch: English Culinary Manuscripts of the Fourteenth Century*, ed. Constance B. Hieatt and Sharon Butler (London: Oxford University Press, 1985).
13. Steven Roud, *The English Year* (London: Penguin, 2008).
14. Lev. 2:14.
15. Lev. 23:14.
16. Josh. 5:11.
17. Ruth 2:14.
18. 1 Sam. 17:17 and 25:18.
19. Talmud Menach. 66a.
20. Edward Brown, *Tassajara Bread Book* (Boulder, CO: Shambhala, 2011).
21. "Lechem Ha'Panim" translates to "Bread of the Face-to-Face Encounter."
22. Orysia Tracz, "Paska and Babka Forever," Orysia's Blog, published April 14, 2012, http://orysia.blogspot.com/2012/04/normal-0-false-false-false-en-ca-x-none.html; and Andre Pur, "Happy Easter with Romanian Traditional Food," Travel Gumbo, published April 21, 2014, http://www.travelgumbo.com/blog/happy-easter-with-romanian-traditional-food.
23. Talmud Ber. 40a. See "Babylonian Talmud: Tractate Berakoth," Come and Hear, http://come-and-hear.com/berakoth/berakoth_40.html.
24. Ruth 2:14.
25. Nihal Kadıoğlu Çevik, "The Pilaf Tradition in Turkish Cuisine," Turkish Cuisine, http://www.turkish-cuisine.org/culinary-culture-202/the-pilaf-tradition-in-turkish-cuisine-226.html.
26. Sophie D. Coe and Michael D. Coe, *The True History of Chocolate* (New York: Thames and Hudson, 2013).
27. "FEAST: Folklore Education & Storytelling for Teachers," Bank Street Graduate School, https://www.bankstreet.edu/graduate-school/professional-resources/feast/.
28. Emiko Davies, "Pastiera Napoletana," Emiko Davies, http://www.emikodavies.com/blog/pastiera-napoletana/.
29. Joyce Goldstein, *Cucina Ebraica: Flavors of the Italian Jewish Kitchen* (San Francisco: Chronicle Books, 1998).

Appendix 1: Baker's Formulas

1. Raymond Calvel, *The Taste of Bread*, trans. Ronald Wirtz (New York: Springer, 2001).
2. Ibid.

Appendix 2: Why Modern Wheat Is Making Us Sick

1. A. Ivarsson et al., "Epidemic of Coeliac Disease in Swedish Children," *Acta Paediatrica* 89, no. 2 (February 2000): 165–171, http://www.ncbi.nlm.nih.gov/pubmed/10709885;.and P. Laurin et al., "Increasing Prevalence of Coeliac Disease in Swedish Children: Influence of Feeding Recommendations, Serological Screening and Small Intestinal Biopsy Activity," *Scandinavian Journal of Gastroenterology* 39, no. 10 (October 2004): 946–952, http://www.ncbi.nlm.nih.gov/pubmed/15513333.
2. "NIH Consensus Development Conference on Celiac Disease," NIH Consensus Development Program, June 28–30, 2004. https://consensus.nih.gov/2004/2004CeliacDisease118html.htm.
3. A. Rubio-Tapia et al., "Increased Prevalence and Mortality in Undiagnosed Celiac Disease," *Gastroenterology* 137, no. 1 (July 2009): 88–93, http://www.ncbi.nlm.nih.gov/pubmed/19362553; and S. Lohi et al., "Increasing Prevalence of Celiac Disease Over Time," *Alimentary Pharmacology & Therapeutics* 26, no. 9 (November 2007): 1,217–1,225, http://www.ncbi.nlm.nih.gov/pubmed/17944736
4. Dr. Hetty van den Broeck, communication with the author.
5. A composite characterization.
6. Abdullah A. Jaradat, "Wheat Landraces: A Mini Review," *Emirates Journal of Food and Agriculture* 25, no. 1 (2013): 20–29, http://www.scopemed.org/?jft=137&ft=15376-43076-1-PB.
7. M. S. Fan et al., "Evidence of Decreasing Mineral Density in Wheat Grain Over the Last 160 Years," *Journal of Trace Elements in Medicine and Biology* 22, no. 4 (2008): 315–324, http://www.ncbi.nlm.nih.gov/pubmed/19013359.
8. L. Dotlačil et al., "How Can Wheat Landraces Contribute to Present Breeding?," *Czech Journal of Genetic Plant Breeding* 46 (2010): S70–S74, http://agriculturejournals.cz/publicFiles/18055.pdf.
9. Lutein may reverse macular degeneration.

10. Hetty C. van den Broeck et al., "Presence of celiac disease epitopes in modern and old hexaploid wheat varieties: wheat breeding may have contributed to increased prevalence of celiac disease," *Theoretical and Applied Genetics* 121 (2010): 1,527–1,539, https://www.ncbi.nlm.nih.gov/pmc/articles/PMC2963738/pdf/122_2010_Article_1408.pdf.

11. A. Sapone et al., "Spectrum of Gluten-Related Disorders: Consensus on New Nomenclature and Classification," *BMC Medicine* 10, no. 13 (7 February 2012), http://www.ncbi.nlm.nih.gov/pubmed/22313950.

12. Personal correspondence.

13. Hetty C. van den Broeck et al., "Presence of celiac disease epitopes in modern and old hexaploid wheat varieties: wheat breeding may have contributed to increased prevalence of celiac disease," *Theoretical and Applied Genetics* 121 (2010): 1,527–1,539, https://www.ncbi.nlm.nih.gov/pmc/articles/PMC2963738/pdf/122_2010_Article_1408.pdf.

Index

Note: Page numbers in *italics* refer to photographs and figures; page numbers followed by *t* refer to tables. Page numbers followed by *ci* refer to the color insert section.

INDEX

negative seed selection, 41
Neolithic and Paleolithic diet, 159–161
nitrogen, 53–54, 61–62, 79–80
no-till approach. *See* reduced and no tillage
Nursit types, 136

oats
    as cover crop, 59
    in crop rotations, 62
    in medieval farming, 142, 144
    seed depth, 69, 76
obesity and obesogens, 8–9
Ohalo II (ancient settlement), 115
on-farm conservation efforts, 37–38, 234
organic farming. *See* landrace grain husbandry
ovens. *See* bread ovens
overseeding, 61

Pacific bluestem landrace wheat, 150
pale malt, 164–65
Palestine
    seed saving in, 37–38
    wheat imports, 20
parched green wheat, 103, 172–74
peas, 59, 60, 62
pedigree breeding, 3
Persian wheat (*Triticum turgidum* subsp.
    carthlicum), *5ci*, 29, 34, 35, 68, 127–28
pesticides, 6, *8*, 45–46
Peter Henderson Seed Company, 152
petroleum industry, 5–6
Pfeiffer, Ehrenfried, 78–79
pharoah wheat. *See* mirabil
phosphorus, 79
pH requirements, 58
plant indicators of soil health, 78–79
planting guidelines. *See* landrace grain husbandry
planting times
    cover crops, 58, 59, 60–61
    factors in, 54–56
    fall vs. spring tillering, 63–64
    moon cycles, 56, 140
    rotation timing, 62
    traditional einkorn farming, 108
Polish wheat (*Triticum turgidum* subsp.
    polonicum), 28, 34, 38–39
Poltavka landrace wheat, 149
positive seed selection, 41

poulard (*Triticum turgidum* subsp. *turgidum*),
    5ci, 28, 34, 38–39, 142, 144
pounding, to remove hulls, 103, 108, 170, 173
powdery mildew, 53
PQ Products, 82
prefermentation, 175, 176
Pringle, Cyrus, 156–57
Progress oats, 157
protein
    in einkorn, 105
    modern vs. landrace wheats, 15, 227–28

quackgrass (*Elymus repens*), 88
querns, trough, 161

ravelled bread. *See* maslin
Ray, John, 150
reaper-binder harvesting, 101
recipes
    art dough, 185
    babka and pashka, 182–83, 202–3
    baker's formulas, 221–24, 222*t*
    Bubliki Russian Bagels, 192–93
    Celebration Bread, 184–85
    Challah, *10–13ci*, 180–82
    Chocolate Babka, 202–3
    Chocolate Biscotti, 204–5
    CR Lawn's Einkorn Ricotta Blintzes, 217–18
    Crust for Quiche and Tortes, 211–12
    dumplings, 197–99
    Easy-Breezy Drop-Dough Kichel, 217
    Einkorn-Almond Milk, 169–170
    Einkorn Bagels, *14ci*, 191
    einkorn chai, 166
    Einkorn Coffee, 168
    Einkorn Kreplach, *15ci*, 197–98
    Einkorn Milk, 170
    Einkorn Sandwich Bread, 187–88
    Einkorn Sourdough Sprout Bread, 186–87
    Einkorn Tabouleh Sprout Salad, 199
    Festive Grain Melange, 196
    frumenty porridge, 171–72
    Gingerbread, 215
    Grain Berry Kvass, *9ci*, 169
    Heirloom Squash Meringue Pie, 209–11
    Horchata, 169–170
    Hrudka Egg Cheese, 184
    Hummus, 200

INDEX

triticale, 60, 62
Triticum spp., 25. *See also specific types*
trough querns, 161
tsiteli doli landrace wheat, 35
Tudor, Nigel, 104
Turkey Red landrace wheat, 152–53
Tusser, Thomas, 143

UG99 rust, 106
Ukrainka landrace wheat, 158
United Nations Food and Agriculture
  Organization (FAO), 4
US Department of Agriculture (USDA)
  1898 plant exploration trip, 153
  1922 bulletin on American wheat varieties,
    145, 152
  SARE program, 4, 35, 38, 45, 104
  seed spacing trials, 73
  warning against poulard and Polish
    wheats, 38–39
US Patent Office, 154–55

Valley Malt, 165
van den Broeck, Hetty, 228, 229
Vavilov, Nikolai, 21, 24, 29, 31, 32, 35, 66,
  147–48, 153, 158
Vavilov's wheat (*Triticum aestivum* subsp.
  vavilovii), 4ci, 32
vernalization requirements, 54
vertical resistance, 46–47
video resources, 234
Vilmorin, Philippe-Victoire, 28, 67–68, 149
Vilmorin seed company, 28, 75, 76

Wadi Haritoun, 8ci, 134
waters, sacred, 124
weather extremes, adapting to, 51, 70
weed management, 22, 25, 38, 78–79, 85–91, 101
wheat rust, 45–46, 106, 145
white babka, 182, 183
white lammas landrace wheat, 150
whole-systems farming, 49–50, 51–52
wild einkorn (*Triticum boeoticum*), 22–23, 27,
  105, 115
wild emmer (*Triticum turgidum* subsp.
  dicoccoides), 2ci, 4ci

in ancient cultures, 115
domestication of, 22–23, 105
mycorrhizal associations, 66
overview of, 19, 27
wild mother wheat. *See* wild emmer
Williams tine weeders, 86–87
winnowing, 102
winter-kill cover crops, 58, 59
winter wheat
  factors in choosing, 54–55, 108
  planting time for, 55–56
  root systems of, 54
  soil fertility considerations, 58
  soybean intercropping with, 61
  tillage for disease prevention, 91
  types of, 30
witchgrass (*Elymus repens*), 88
World Wide Wheat Company, 6

Xinchan Rice (*Triticum aestivum* subsp.
  petropavlovskyi), 32

yeast, 10, 177. *See also* sourdough starter
Yehudit (shopkeeper), 148
yellow babka, 182, 183
yellow lammas landrace wheat, 150
yellow mustard, 59
yields
  fall vs. spring tillering, 63–64
  Green Revolution accomplishments, 5
  modern vs. landrace wheats, 4, 227
  mycorrhizae benefits, 65–66
  SARE-funded trials, 38
  seed spacing effects on, 69–75
Yusef (farmer), 66

zanduri (*Triticum turgidum* subsp. *timopheevii*),
  5ci, 28, 34, 62
Zarathustra, 126–27
zbizhzhia (Ukrainian word for grain), 184
Zea spp., 25
Zhukovskyi wheat (*Triticum aestivum* subsp.
  zhukovskyi), 32, 34
Zkukovskii, Mikhailovich, 27
Zook, Samuel, 152
Zoroastrian traditions, 126–28, 132

About the Author

When Eli Rogosa, a food and farming anthropologist, agronomist and seed saver, worked with farmers in the Fertile Crescent, she discovered a treasure of landrace food crops thriving in the stifling Middle Eastern heat without irrigation or chemical intervention. Fascinated, she soon learned that these robust wheats are on verge of extinction, and founded the Heritage Grain Conservancy to collect the rare wheats before they were lost to the world. Eli was funded by the European Union to be a research partner

Shulamit Falik

with European gene banks to collect landrace wheats. Today Rogosa grows hundreds of rare wheats and manages an artisan einkorn bakery on her biodiversity farm in Western Massachusetts. She farms with her companion, CR Lawn, and has two sons, Noah and Ezra.

the politics and practice of sustainable living

CHELSEA GREEN PUBLISHING

Chelsea Green Publishing sees books as tools for effecting cultural change and seeks to empower citizens to participate in reclaiming our global commons and become its impassioned stewards. If you enjoyed reading *Restoring Heritage Grains*, please consider these other great books related to food and agriculture.

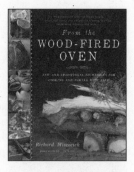

THE NEW BREAD BASKET
How the New Crop of Grain Growers, Plant Breeders, Millers, Maltsters, Bakers, Brewers, and Local Food Activists Are Redefining Our Daily Loaf
AMY HALLORAN
9781603585675
Paperback • $17.95

FROM THE WOOD-FIRED OVEN
New and Traditional Techniques for Cooking and Baking with Fire
RICHARD MISCOVICH
9781603583282
Hardcover • $44.95

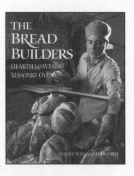

HOME BAKED
Nordic Recipes and Techniques for Organic Bread and Pastry
HANNE RISGAARD
9781603584302
Hardcover • $39.95

THE BREAD BUILDERS
Hearth Loaves and Masonry Ovens
DANIEL WING and ALAN SCOTT
9781890132057
Paperback • $35.00

CHELSEA GREEN PUBLISHING
the politics and practice of sustainable living

For more information or to request a catalog, visit **www.chelseagreen.com** or call toll-free **(800) 639-4099**.